THE FOURTH KINGDOM

THE FOURTH KINGDOM

William J. Sauber

First Edition

AQUARI
CORPORATION

Box 1966
Midland, Michigan 48640

ACKNOWLEDGMENTS

Editorial Assistance
        E. G. Baldes
        E. N. Brandt
        Paul N. Sutton
        Alice B. Ziegler

Illustrations
        Gary Gustaf

Copyright © 1975 by William J. Sauber
All rights reserved.
Published by AQUARI Corporation, Box 1966, Midland, Michigan 48640
International Standard Book Number: 0-916204-00-6
Library of Congress Catalog Card Number: 75-27461
Printed in U.S.A.
First Edition
First Printing, November 1975

The efforts, ideas, counsel, guidance and encouragement from a great many people, some known and some unknown to me, made this book possible. If you who are reading this are one of those people, and I am certain you will know, please accept my sincerest appreciation and thanks.

BILL SAUBER
August 31, 1975

CONTENTS

Chapter 1	Omega—The End of the Beginning	1
Chapter 2	The Beginning	12
Chapter 3	The Flowering of Life	20
Chapter 4	A Creature of Promise	35
Chapter 5	Evolution of the Fourth Kingdom	42
Chapter 6	The Ark	66
Chapter 7	Plurality of Purpose	76
	Agriculture	82
	Culture	83
	Sexuality and Society	84
	Miniaturization	91
	Science	93
	Communications	96
	Race and Adaptability	101
	Nuclear Fusion	104
	Employment and Social Welfare	106
	Plurality's Arrow	107
Afterthoughts		109
Bibliography		116
Illustrations		
	Simplified Tree of Animal Life	27
	Evolutionary Tree of the Fourth Kingdom	61

Chapter 1

OMEGA – *The End of the Beginning*

> No other object in the universe is as important to man as the sun, the central fire upon which depend all life on the earth and any life there may be elsewhere in the solar system.
>
> David Bergamini,
> *The Universe*[1]

Deep inside a Colorado mountain, General Walter Sanderson was finishing his shift as commander of the Air Force Nuclear Missile Control Center. Sanderson completed briefing his relief commander, and then left the rock cavern and its array of electronic consoles which controlled most of the armed nuclear missiles in the United States.

The General stepped into an elevator which rose silently in a shaft cut from solid rock. It took only twenty seconds to reach the surface. He walked quickly through a steel-lined tunnel to the squat, concrete security building which stood guard over the tunnel entrance. He completed the ritual of checking out through the Security Officer and then stepped out a door into the bright afternoon sunlight.

As his eyes adjusted to the light, General Sanderson's vision was filled with the magnificent mountains and deep blue sky of the Colorado high country. He was relieved as usual

---

[1] Bergamini, David, *The Universe* (New York: Time-Life Books, c. 1962, 1969, Time Inc.), p. 95.

to turn his command over to the next shift—that awesome control at his fingertips to either destroy, or spare from destruction, great areas of the earth. As he looked at the natural grandeur about him, he thought briefly about the power that mankind held over nature. From primitive beginnings, where man competed with all nature's creatures for survival, we had now developed the power to control all the earth.

The General's career in the Air Force, especially the last five years of it, had forced him to lead a well-ordered life, so his actions were usually predictable. Today, as was his habit while walking toward his car, he pulled a heavily embossed cigar case from the pocket of his jacket.

Smoking was forbidden in the command room, so he always took a great deal of pleasure in lighting up a fresh cigar as he walked the hundred yards to the parking area. Almost unconsciously, he selected a cigar from the case and brought out his lighter. The snap and the short rise of flame from the gold-plated lighter would mark the end of General Walter Sanderson's military career and the great power he commanded. He would be dead in a matter of minutes.

Long before the General had started for the elevator deep within the mountain, a chain of events had started deep within the sun. The non-hydrogen ash in the sun's core had become so voluminous that the sun could no longer continue to burn as a stable, gentle star. The sun had to start making adjustments called for by physical law.

Like the usual signs of upset from a heaviness in the stomach, the sun would develop "solar hiccups" from deep inside. But even a "hiccup" on a solar scale could be devastating to the planet earth orbiting so near the sun's usually benevolent power.

The upset in the core caused a huge pulse of energy to rip through the heavy midsection of the sun. On this particular day, the rapidly expanding firestorm of energy burst out through the sun's chromosphere in a geyser-like eruption.

From earth, the burst looked something like a solar flare, except that it was enormous.

Raw, unchecked radiation of indescribable power beamed into space from the sun. The ozone layer around the earth, which protects life from most of the sun's high energy radiation, was mutilated by the first wave of energy to reach the earth's outer atmosphere. Deadly radiation then poured onto the surface of the earth like a midsummer cloudburst.

The General saw the Colorado sky turn from a deep blue to an eerie grayish-green, then flicker like a faulty flourescent tube. For an instant, the General thought the impossible had happened—an enemy nuclear attack had commenced and his forces' warning system had not detected the incoming missiles. He quickly noted the phenomenon was not like a bomb burst. There were waves of color and light across the total horizon. He then felt an irritating tingle throughout his body and a burning feeling on his skin. He started to run for the security building. He felt his legs collapsing under him and felt his strength and consciousness rapidly ebbing. He then knew he was the victim of some universal disaster he did not understand.

General Sanderson lost consciousness and fell to the ground dying from the massive radiation damage to the tissues of his vital organs.

At this same moment, thousands of miles away, Ms. Felicia Baker, a mainland tourist, was getting ready to leave her pew along the back wall of the old stone missionary church on the island of Oahu, Hawaii. As church services ended, Ms. Baker thought that indeed mankind had dominion over all the earth. In less than an hour, she would receive sustenance from the fruit of plants, the eggs of fowl, and the flesh of animal—all to be enjoyed in the garden paradise that was Hawaii.

People were already falling in the street just outside the church. If Ms. Baker had stayed in her seat to contemplate for just three more minutes, she would have been one of the fortunate who survived the rain of deadly rays that lasted only six minutes. The thick stone wall of the church would have shaded her from the radiation.

But Felicia was thinking of the delayed breakfast she would savor, and she hurried out the church door and down the worn steps. She was out in the street before she noticed a number of things all at once. The light outdoors was in fact strange, which she at first thought was a problem of her eyes adjusting to bright daylight. Her body was tingling. Her skin felt like an attack of prickly heat. There were people lying about on the street. She quickly weakened and fell prostrate in the street less than twenty-five yards from the stone church.

Ms. Felicia Baker did not have time to recognize that she was just another one of the creatures of nature's domain, so vulnerable to destruction by natural forces.

Of course, General Sanderson and Felicia Baker and the events of their fate never did exist, except in the author's imagination. But these fictional characters do dramatize a reality that confronts all life on earth. Humanity came face to face with this reality when we first peered out beyond the boundaries of earth. Out there we found hurtling comets, exploding stars, colliding galaxies and powerful quasars. Life on earth then seemed like a fragile flower growing in a lovely meadow, but with brutal giants stamping about in the surrounding wilderness.

Man was haunted again by that persistent legend of Noah which reminds him that all life can be endangered by natural forces. However, forces beyond the earth present far greater risks than a simple flood, since most classes of plants and animals contain species which can survive very well in a watery world. But life could not survive the searing heat, intense cold or deadly radiation so common to our universe.

As astronomers continue to look at the hundreds of billions of stars in the universe, they report further ominous findings. Stars the size of our sun, and larger, do not age in a steady, graceful manner. The largest stars age the fastest and end their existence with a cosmic explosion that would pulverize, if not vaporize, any nearby planets. Stars the size of our sun do not explode, but they reach a period of instability and in time change into red or orange giant stars as they near old age.

Astrophysicists have worked out a generally accepted version of how a star like our sun is expected to behave. During its early years and through middle age the sun is energized by the nuclear burning or fusion of hydrogen taking place in its core. The fusion of hydrogen produces helium as a by-product, which at first acts as an inert ash gathering in the sun's core. However, when the mass of helium becomes large enough, a hotter, more energetic reaction will start in the sun. Some astrophysicists feel the faster reaction will start when the helium content of the sun reaches about 15 percent. At that time the sun will no longer be a middle-sized yellow star, but will swell to a red giant, remaining so for a billion years.

Scientific authorities tell us that the life of the sun as a stable yellow star is perhaps ten billion years; and since it has been burning for only five billion years, there may be another three to five billion years of tranquility left. However, there are some uncertainties in this prediction.

The ten-billion-year period as a hydrogen burning star *assumes* that our sun condensed from a primordial gas cloud consisting almost completely of hydrogen. Yet, the presence of heavy metals in the sun, and certainly the presence of heavy metals in the sun's planets, say our sun condensed from the ashes and gases of *other* stars that lived and died. So, the sun is likely a third or fourth or more generation star that began life with an already significant, but unknown quantity of helium and other "ashes."

We think we know how much helium is being produced in the sun: 652 million tons every second. We don't know how much was in the sun to start with. So any estimates of the number of years we can expect to live in the radiance of a peaceful star are very rough at best.

Other stars which have evolved to red giants have been seen only from great distances through telescopes. Astronomers cannot see the smaller changes in a star as it proceeds toward being a red giant. Yet, these small changes are on a scale that becomes dreadfully important, if that star happens to be your sun and you live on one of the inner planets. The surface of the earth is so affected by the radiance of the sun that in reality we are living in the outer body of the sun. As expressed in the Beatles' popular verse, "We all live in a yellow submarine."

We should sense and may even suffer disastrous effects from as yet unknown changes in the sun—for example, deadly radiation flashes during periods of sun instability—before a red giant emerges in the sky.

Long term changes in the sun's radiation are reflected by changes in the climate of the earth. In relatively recent geological times the earth has been cooling, indicating a lower radiant output from the sun. Then, in the most recent geological age, the earth's climate has been fluctuating between ice ages and warming spells. The most reasoned analyses of the ice ages have attributed them to variations in the sun's radiation. Thus, there is evidence of a flickering or radiating instability which must give us pause, for a reduction or fluctuation of radiation could easily be attributed to a depletion of hydrogen, the primary fuel, and a gathering of inert "ash" in the sun's burning chamber. Thus, the sun could be expected to cool even further or become more erratic prior to the onset of a hotter fusion reaction.

There should be other changes evident to us here on earth of any growing unsteadiness in the sun. The change in numbers, size and frequency of sun spots may give us clues. We regularly note solar flares and prominences and emissions of particles from the sun. These may be normal behavior or may be signs of instability. We don't know yet.

The words of an anonymous Hebrew prophet of 450 B.C., recorded in the *Bible's Book of Malachi,* portray tragedy for mankind:

> For lo, the day is coming, blazing like an oven
> When all the proud and all evil-doers will be stubble.
> And the day that is coming will set them on fire.[2]

Scientists too, are equally certain that the living earth will end in fire. A day is coming, say the astrophysicists, when our sun will swell to become a red giant star. The earth's surface will then be a holocaust in the 1000° F heat. The oceans will boil away, and all life on the earth's surface will be reduced to ashes. Civilizations on earth will be forced to exist underground like huge ant colonies. But there can be no hope of survival since the sun will remain a giant red star for a billion years. Man and his cities can burrow only so deep into the earth before reaching intolerable heat coming up from the earth's molten interior.

The underground environment will in time become unbearable, as heat from earth's scorched surface flows downward and heat from the earth's core flows upward. Life on earth will then slowly perish in a suffocating hell.

---

[2] *Malachi* 2, 19.

A residue of hot ashes is an unreasonable fate for the boundless wonders of life. A flaming trap is an improbable ending for creatures that have survived over billions of years under the most difficult conditions and circumstances.

We see purposeful movement—mobility—as the most important reason for life's surviving and flourishing in every conceivable place. Birds and insects travel throughout the ocean of air surrounding earth; fish move through all the seas; fleet-footed animals move swiftly over the land. Even stationary plants find mobility through their seeds, which serve as tiny life insurance packages to be carried by the winds, the waters, and the movement of animals. In this power of movement, life has protected itself from life-destroying disasters—fire, earthquakes, volcanoes, and disease plagues—which may strike one meadow or one forest or one continent.

Only a new kind of biological mobility could prevent the total destruction of a life system imprisoned on one planet and dependent on one star. The three classical kingdoms, animal, vegetable (plants) or mineral, offer no means of escape. The cells of plants and animals consist of delicate chemicals suspended in water. It is also unlikely animals could evolve enough muscle power to escape earth's gravity; if they could, their tissues would be destroyed in the chill and vacuum of outer space. Minerals could survive the temperatures and vacuum of space and could take great physical stress. Minerals, though, are lifeless and have no power of movement.

Therefore a new biological phenomenon evolved, a *fourth kingdom* of creations which were mineral in substance, but lifelike in form and lifelike in their function. The beginning of that phenomenon was the coming of man. From the purposeful workings of the boundless creative power in the life system, a new animal, man, evolved who was endowed with a far greater excess of brain cells than was needed for animal survival. This excess of biological brainpower supported the building of a civilization and technology. The fourth kingdom could emerge from only a highly ordered civilization.

The living and dying of billions of simple creatures over billions of years was required to evolve a bird or a mammal from the stuff of life. Similarly, the "living and dying" of millions of mechanical creatures—chariots, ships, locomotives, cars, airplanes—was needed to evolve a space vehicle to carry life beyond this planet. Machines have evolved which give hope to the most critical need of life—the mobility to survive a disaster striking the whole planet. The evolution of machines showed a remarkable parallel to animal evolution, but this was not surprising. Machines evolved from brainpower of the animal kingdom residing in man. The evolution of the fourth kingdom will continue until space arks come into being which can carry earth's life to new planets.

If we look carefully at the creativity in nature, we see that man and the fourth kingdom are not spectacular or even unusual. The simple slime mold, an organism with the forbidding name of *Dictyostelium discoideum,* lives normally as a single amoeba-like cell, a formless glob of matter slithering about capturing bacteria like a tiny hunting animal. Yet, when the bacterial food supply begins to run out, these one-celled animals cluster together to form a colony of about fifty thousand. The colony then proceeds to erect a small tower, like a flower stem with a bulge at the top, where new seeds of life are created and launched into the world. This launch tower is formed from the bodies of these tiny animals in an elaborate cooperative effort involving the whole colony.

In a way no more wondrous, only more grandiose, humanity has paralleled this simple amoeba. Mankind evolved from scattered hunting tribes to civilizations involving cooperative efforts of millions of individuals. Then, thousands of people, each with billions of brain cells, working in purposeful harmony, built launch towers and rockets, and found a way to launch the seeds of life to the universe. The timeless evolutionary plan, to assure the perpetuation of life, has simply repeated itself in man and the fourth kingdom.

There is no need to plead the cause of man's destiny in space, for that will be fulfilled whether or not books are written, or wars fought, over the subject. The important objectives of life are met sooner or later by a purposeful and often painful evolutionary process. Nature's needs are met regardless of what the members of one species of life may do.

Countless species of plants and animals have perished over the ages because they were unable to adapt to nature's evolutionary objectives. The same rules apply to human societies, nations, or civilizations. People who did not embrace the fourth kingdom have perished, or are living in the backwaters of the earth, dominated by the people with advanced technology.

There can be only a bleak future for a society that brings its technology to maturity but does not satisfy life's desire to pervade the universe. A negligent culture or nation will somehow be swept aside. The people will turn against their leaders and each other; their will to succeed will stagnate and their governments and industries will degenerate. The creative power of our great life system will make room for those who will use their technology to assure the survival of life.

A worldwide effort to carry life to the universe would save nations and their people from considerable anguish. If the nations of the earth work together to solve the enormous challenge of a workable space ark, lasting world peace would be a probable result. The required creativity and competition would absorb human energies normally reserved for war. Perhaps civilization will then be spared the near certainty of a man-made nuclear holocaust in a territorial war. A peaceful, constructive, multi-national effort would rain down technological benefits over all the earth bringing a better life for all.

Science and religious philosophy both support the belief that man and the fourth kingdom will find eternity for life in the heavens. Astronomers long ago concluded a great many of the stars and their planets in our galaxy could support life. Thus, in the far heavens, life can find salvation after the earth becomes an uninhabitable place.

The *Book of Isaiah* is widely quoted in Western civilization. *Isaiah* tells of the dilemma of a person or a nation faced with decisions which lead to destruction or salvation, punishment or reward. The words of *Isaiah* also foretell a punishment by fire:

> Lo, the Lord shall come in fire,
>   his chariots like the whirlwind,
> To wreak his wrath with burning heat
>   and his punishment with fiery flames.[3]

Isaiah speaks of the promise of rewards for those who follow the Lord's way. There are two references in the last chapters of *Isaiah* to the promise of "new heavens and a new earth." The oracle who wrote those words centuries ago did not know exactly where there was a new earth in new heavens; its presence was guaranteed by God. By definition, an oracle does not speak from the precision of facts; rather, he speaks from the heart of the Power of Creation. In the closing verses of *Isaiah* are words which continue to ring in our ears much like a church bell's last peal, hanging in the still air before a storm:

> As the new heavens and the new earth
>   which I will make
> Shall endure before me, says the Lord,
>   so shall your race and your name endure.[4]

---

[3] Isaiah 66, 15.
[4] Isaiah 66, 22.

Chapter 2

THE BEGINNING

> Energy will remain in some sense the lord and giver of life, a reality transcending our mathematical descriptions. Its nature lies at the heart of the mystery of our existence as animate beings in an inanimate universe.
>
> Freeman J. Dyson,
> "Energy in the Universe"[1]

Before the beginning, there was no earth and there was no life. A dust cloud drifted among the stars where the earth had yet to take form. The absence of life was a dark and unimaginable void. For without life there could be neither earth by word, nor any words, nor sights, nor sounds, but only blind, senseless energy flowing through an unconscious universe.

The planet earth began in the Milky Way galaxy, one of countless galaxies scattered throughout our universe. The Milky Way is an immense island of stars, gas, and dust, separated by billions upon billions of miles of black void from its nearest galaxy neighbor.

The galaxy of our birth is in the form of a disc with spiral arms, resembling a fireworks pinwheel turning slowly in space. The hub of the pinwheel is a reddish-yellow globe and the spiral arms sparkle with blue and yellow light from the nuclear fires of over 100 billion stars.

Clouds of dust billow among the spiral arms, shading some of the

---
[1] Dyson, Freeman J., "Energy in the Universe," *Scientific American*, September, 1971. (c. 1971, Scientific American, Inc.), p. 51.

stars and their light. The dust was manufactured in the nuclear furnaces of many generations of stars. Some stars ended their existence by exploding and scattering stardust for billions of miles.

It was probably the gentle pressure of light from distant stars that caused a part of one dust cloud to begin to take form about five billion years ago. The force of gravity did the rest. Dust and gas rushed toward new centers of gravity.

As the cosmic cloud condensed, a large central ball of matter formed and gathered up most of the dust and gas. Finally the crushing pressures and the temperatures at the center of the ball ignited a nuclear reaction. As the mass of the orb increased, so did the intensity of the nuclear reaction until a new star—our sun—was born. The earth and the other planets came from the cold dust left in space as the sun formed. Had not enough of the dust cloud been apportioned to our sun, its nuclear fires would not have ignited, and our solar system would have remained dark.

It would be hard to imagine the earth without light. After the first billion years or so, the surface of the earth would become as cold and as dark as the voids of space. There would be no water to flow; no wind would blow. The feeble light of stars would be faintly reflected from the icy surface. There would be no eyes to see earth. The earth would be lifeless, its darkened hulk invisible from the vacuum of space surrounding it.

Darkness, however, was not to be earth's fate. Earth had taken form near a modest sized yellow star, shining from one of the galaxy's spiral arms some twenty thousand light years distant from the center of the Milky Way. The earth received just enough of the warm radiance of the sun so that most of the planet's water stayed in liquid form. The oceans and clouds of earth glowed in the sunlight. Earth was a blue and white jewel against the blackness of space. This blue planet, this water planet, was a wonder in our solar system and a wonder in all the Milky Way galaxy.

Most of us view matter as solid and substantial, and energy as fleeting and invisible. Yet, if we are to understand how life came to be on this planet, we will have to face a mind-jolting truth that physicists have known for years: Matter and energy are interchangeable!

Matter consists of bundles of energy waves so powerful and so stable that we call them solid. "Something solid," becomes to a physicist simply energy at rest. The physicist insults our instincts further by measuring solid matter in terms of million electron volts, a unit of energy.

We find in this nuclear age that matter is energy in harness. The atomic bomb and the hydrogen bomb will not let us forget the enormous energies which escape when we disturb matter and break apart a small bit of that energy bundle. In a nuclear reaction, matter simply "disappears" and becomes energy.

The happenings in our high-energy particle accelerators, our "atom smashers," are equally startling. At Stanford in California, and at Long Island's Brookhaven National Laboratory, *physicists turn energy into matter on a routine basis.*

Physicists, looking deeply into the structure of atoms, find that matter and energy are hardly distinguishable. When subjected to intense forces, matter and energy shimmer back and forth across the boundary separating them.

This oneness of energy and matter makes the phenomenon of life possible on this planet earth. For life is something that exists only at the very boundary between energy and matter; life exists in a delicate balance between the two.

Radiant energy from the sun was to be the most important ingredient for our living earth. We know the content of the sun's rays as light and heat. This is what our senses tell us. We measure "wave lengths" and we see the "colors" in sunlight. However, the sun's rays are a blend of radiant energies so complex that we may never understand their significance.

The sun's radiance comes from a nuclear process called hydrogen-fusion. In the core of the sun, 657 million tons of hydrogen are converted into 652½ million tons of helium each and every second. The missing 4½ million tons of matter are annihilated and converted into pure energy. As the Aztecs, with misguided zeal, tore the hearts from thousands of living victims as sacrifices to the sun, so in the sun the very heart is torn from matter, destroying it forever, to produce beams of energy and the promise of life for the planet earth.

The cosmic setting for the beginning of life was dramatic; the physical reality of the scene was poetic. In the core of the sun, matter was being obliterated forever, in order to send radiant energy to earth, a sphere of ashes from long dead stars. Something had to come from the workings of these cosmic forces, from the sacrifice of so many stars and so much matter; something more than heated rocks had to result.

We know that life was to come to this special planet and its sun. However, those billions of years ago, the phenomenon of life arising from this barren planet was unimaginable. The seeds of life would not be detected among the dead elements of earth. Sunlight was there but it was too transient, too fugitive, to provide a living form. Life would have to come from some miraculous blend of earth's dead matter and sunlight.

The whole process of life was too incredibly complex to contemplate in the beginning. However, there was the power of creation and there was time, billions of years of time.

In the 1950's science was to have its first hint of the power of creative chemistry contained in light. In laboratories, investigators mixed water vapor, hydrogen, ammonia, and methane, the most probable gases present in the early atmosphere of earth. The gases were exposed to ultraviolet light, one of the portions of sunlight. The light caused amino acids and other chemical building blocks of life to appear in the flask.

With some amazement, the investigators had difficulty explaining this "natural chemical synthesis of the stuff of life." Had there been a philosopher among them, an explanation would have been obvious. For here was demonstrated the creative power of light. Here was light gathering together dead molecules of matter, infusing their atoms with energy and rearranging the molecules to a form through which energy could flow and become "lifelike." In that laboratory was a recreation of the beginning of the living earth.

Had it not been for the abundance of water on the planet earth, the creative groping of light energy would have been in vain. Without the oceans to moderate the excessive heat of the day and the deadening chill of the night, the first delicate chemicals would have been destroyed. The gentle, warm oceans of the world were natural incubators. Earth's great oceans were also rich in chemical substances, mixing in the warm currents. The waters of the earth were a plowed and fertilized barren field, awaiting the seeds of life.

For untold millions of years amino acids, proteins, and enzymes—chemicals of life—formed in the waters of the world by the action of sunlight. As lands began to rise above the oceans, becoming the early continents, chemical experiments went on in countless millions of sunlit pools. During these days of genesis other natural forces were at work. The moon and the tides filled and flushed seashore pools where life's chemicals were forming. The power of lightning, the heat of volcanoes, and perhaps even the fiery entry of meteorites into the atmosphere, made the more difficult chemistry possible.

From the billions of experiments with the chemistry of light, there occurred a critical event. Somewhere and sometime during a span of millions of years, molecules developed which could reproduce themselves rapidly. Lifelike only in their ability to reproduce, pre-life chemicals multiplied at an enormously accelerated rate. The pursuit of life stepped up its pace perhaps a billion-fold.

Over more millions of years, the pre-life chemicals continued to grow in organization and complexity in the sunlit waters. Then, the most important event in all of life's history took place, about four billion years ago. The first living cells emerged as complete and independent units of life. Simple cells then evolved to advanced cells which were miniature entities of all life's functions. They could pulsate and move, react to stimuli, grow, protect themselves, and most importantly, reproduce themselves.

The living cell became a tiny city with its own social and industrial activity. Directed by a control center, the nucleus, a cell was capable of turning raw energy into power for work. The cell could also do its own manufacturing and storing of chemicals for growth and power.

A study reported in *The Cell,* of Time-Life's Science Library, is worth repeating. In order to duplicate the chemical reactions which take place in one cell, David Garfinkel of the University of Pennsylvania, used an electronic computer. A computer can reduce days of hand calculations to minutes by the use of electronic wizardry. Yet, assuming only twenty-two chemical reactions in the cell as a basis for his experiment, *Garfinkel needed nine hours of computer time to simulate the chemistry taking place in a living cell in only fifty seconds.* A real cell carries on up to two thousand chemical reactions simultaneously, not just twenty-two. The biological cell, millions of times smaller than our modern computers, runs and regulates these reactions constantly.

How many times have we scorned the existence of a lowly creature, a worm, or a mosquito, or a blade of grass? How often have we considered the simple creatures of nature insignificant compared to our own works, our architectural edifices or our mechanical creations? Yet, with all of the scientific and engineering talent that we could put together on the whole earth, we could not construct something as complex as one single, living, working cell. Dr. Albert Schweitzer in his later years came to appreciate and to revere the wonder of life's simplest creatures; he could not and would not kill even an ant that came into his office or home.

Science had long been curious about the makeup of a cell, for within its intricate structure was the boundary between life and non-life. The cell contained the mystery of our living existence. The search for life's secrets became infinitely more exciting with the perfection of the electron microscope. Magnifications up to a million-to-one revealed new cell parts and their workings for the first time. With heart throbbing excitement, we peered into the very heart of a living cell and into the very secrets of life. However, we did not find a boundary marking living existence, but rather bits of matter in a frenzy of motion. We found that investigating a living cell is like examining an ocean wave approach the shore; we can see the form of the wave, but if we peer beneath it, we see only a chaos of sand and water and pieces of debris.

We wonder if the biologists who looked into the heart of a cell ever talked to the physicists who looked into the heart of matter. The physicists who looked deeper into the atom found only a chaos

THE BEGINNING 19

of forces where matter and energy could not be clearly separated one from the other. As Tom Alexander reported in *Fortune* in 1968:

> No longer, in fact, could an atom or a particle—the very matter that constitutes our reality—be viewed as something that endures as anything more than an act of continuous creation, like a cloud that retains its position, shape, and identity in the fierce wind surging over a mountain peak by continuously melting away downstream and being created upstream. Some thirty years ago, the Irish poet, Oliver St. John Gogarty, coined an apt description of what the physicists now seem to be seeing: 'Chaos contracted to intricate form.' [2]

In the living cell the delicate, intricate phenomenon of life is also a chaos of motion where starlight blends with atoms; where energy becomes form; where lifeless matter becomes infused with energy to pulsate, to move, to feel, to become alive. It is in the living cell where non-being becomes being.

In a far corner of the universe, the miracle of life had appeared. Starlight had found meaning in the image of a living cell, created from among the fertile elements of a blue planet circling in the third orbit around an inconspicuous yellow star. Was this a one-in-a-hundred billion circumstance? Was our sun the only star to find life in the whole of this island universe, the Milky Way galaxy?

---

[2] Alexander, Tom, "The Shimmery New Image of Matter," *Fortune*, June 1, 1968, (c. 1968, Time, Inc.) p. 127.

Chapter 3

THE FLOWERING OF LIFE

> In the mutual reinforcement of these two still opposed powers, in the conjunction of reason and mysticism, the human spirit is destined, by the very nature of its development, to find the uttermost degree of its penetration and the maximum of its vital force.
>
> Teilhard de Chardin,
> *The Phenomenon of Man* [1]

Arguments have been raging for years about how life and mankind came to be on earth. The traditional religious belief is that all life was created in a few days at the beginning. The evolutionists believe that life came about on earth as a cosmic accident, that different forms of life appeared through the working of blind natural forces. Life advanced, says the evolutionist, by accident or chance, assisted by selection or "survival of the fittest" as all life's creatures competed for territory and food. These opposed views have met head-on in the churches, in the schools, and in the courts as well. People have lost their heads or their jobs over the emotions caused by these matters.

The case for instant creation came from an early interpretation of the *Bible's Genesis*. All life was created in a few days, said the church's interpreters. This very literal meaning of *Genesis* grew in

---

[1] Teilhard de Chardin, Pierre, *The Phenomenon of Man*, (New York, Harper and Row, 1965; c. 1959 William Collins Sons & Co., London, and Harper & Row, Publishers Inc., New York), p. 285.

strength during the Middle Ages, and by 1650 Archbishop James Ussher of Ireland stated that the moment of creation took place in the year 4004 B.C. This date became inserted in the margins of authorized versions of the *Bible* as the moment of life's creation.

Studies of the natural sciences grew rapidly in the century after Archbishop Ussher's proclamation. Some inquisitive individuals began picking around in rocks which were millions of years old, and in these rocks they found evidence of life—remains of creatures which no longer existed. Evidence seemed to show that life in earlier times had been much simpler in form and had progressed from simple creatures to the more complex creatures. And it all seemed to have taken place by a long, slow process of evolution. A deepening shadow of doubt was cast over Ussher's pronouncement.

Evidence for the process of evolution became overwhelming. People studying the evolutionary process shouted about, for all to hear, that the religious teachings had preached ignorance and scientific fraud. Darwin's book, *The Origin of Species,* became a cornerstone of the new school of evolutionary science. Scientists with their new vision even attacked the divinity of man with righteous outrage. Man was not a creature apart from nature; except for his larger brain, man was similar to the rest of the animals and nearly identical physically to the apes and chimpanzees. Man was an intimate part of the rest of the life on earth and indeed had evolved from the animal kingdom.

But in their zeal to expose centuries of ignorance, the evolutionists overlooked the possibility of a creative power. Perhaps they thought that if the religious teachings were so wrong on the subject of an orderly, evolutionary development of life, they must be wrong about the power of creation as well. And thus began a never ending battle of words. The conservative religious believer accepted creation, but would not accept that evolution was the way creation worked. The narrow scientist knew that evolution was the means by which life progressed, but would not accept that the workings of evolution were guided by a creative power. Few wanted to move to the obvious middle-ground of compromise in this argument.

One man stepped into the middle of the debate over thirty years ago. His name was Teilhard de Chardin—a distinguished scientist and a Jesuit priest. When he wrote his now famous book, *The Phenomenon of Man*, he was distressed by the opposing views of religion and science. Much to the dismay of his fellow churchmen, Teilhard recognized the workings of evolution in the development of life and said:

> *Like all things* in a universe in which time is definitely established as a *fourth dimension,* life is, and only can be, a reality of evolutionary nature and dimension..... To shake our belief now in the reality of biogenesis [*de Chardin's term for biological evolution*], it would be necessary to uproot the tree of life and undermine the entire structure of the world.[2]

Yet, Teilhard recognized evolution only as the mechanics by which creation reaches life's many goals. For the evolutionist who observes the many forms of creatures and concludes life's progress came about by accident of numbers, Teilhard had this to say:

> ... we find the fundamental technique of *groping,* the specific and invincible weapon of all expanding multitudes. This groping strangely combines the blind fantasy of numbers with the precise orientation of a specific target. It would be a mistake to see it as mere chance. Groping is *directed chance.* It means pervading everything so as to try everything, and trying everything so as to find everything. Surely in the last resort it is precisely to develop this procedure (always increasing in size and cost in proportion as it spreads) that nature has had recourse to profusion.[3]

Recognizing a creative power in our life system is natural to those whose religious training and faith have influenced their lives. Yet, faith alone is not enough for those whose training has been dominated by the scientific process. The Darwinian dogma, accepted for

---
[2] Ibid., p. 140
[3] Ibid., p. 110

a hundred years by most of the scientific community, leaves little room for the presence of an active creative force.

However, the proof that there is a power of creation has been uncovered by the scientific process itself. A common creation is based on the fact that all life's images are the result of electromagnetic energy coming from the sun. Thus, all life is as one, connected through the life-giving essence of sunlight which flows through the whole of the living earth. All life is like an enormous television production which uses an infinite number of channels where life's images are created and where they perform in an endless assortment of programs. Technology mimics the miracle of life in a very crude way by re-arranging electromagnetic energy to create television images on the screens in our homes.

A single living cell exists for the same reason as a television image. Thousands, or perhaps millions, of tiny bits of energy converge and coincide within a microscopic speck of space. Systematic arrangement of these bits of energy makes the living cell function. The higher animals function by further arrangements of billions of these cells.

Living events can therefore be the purposeful arrangements of the same energy. The very existence of the living cell says that multitudes of events will be arranged throughout all of the life system. Events can be arranged across, between and among all life's creatures. The total network of life can thus be programmed and tuned by evolutionary change to solve new problems.

Who or what one calls this creative power is a matter of choice. Our religions use "God," the "Holy Spirit" or the "Great Spirit" interchangeably. We talk of "Mother Nature" and "Creation." *But to call it all "accident-of-chance" is to risk being accused of not doing our homework on the ever deepening findings of science itself.* Dr. Albert Einstein came to understand the physical laws of matter and the universe as well as anyone of his time. Even though Einstein perceived reality far more deeply than most of us, his public statements left no doubt that he believed in a creative force, a higher intelligence, operating in nature and the universe.

Darwin observed only the gross physical characteristics of life's creatures. His work was done long before the invention of the electron microscope and long before the real coming of the sciences of biology and biochemistry. He could not have imagined the miraculous operation of one living cell. He did not know of the coordinated complexity of billions of such cells making up the higher forms of life. *Were Darwin to be alive today and to understand only a fraction of what the new frontiers of the life sciences are learning, he might have included the power of creation as the foundation of his evolutionary theory.*

The flowering of life can take on new and beautiful meanings by our accepting the presence of a creative power working through the evolutionary process. Life can then be seen as a force yearning for more awareness as it proceeds from lifeless chemicals to living cells. In the miracle of a bird, blind energy from the sun finds far more meaning than in the heat of a rock.

A primary driving force of nature is to assure the existence of life wherever it is possible. Therefore, life must have an ability to move around and disperse itself to every part of the planet. So in addition to awareness, there is an evolutionary force to develop creatures that have maximum freedom of movement. It is with this understanding that we follow the development of life from single cells floating in the ocean to the rich and varied life system which covered the earth before the coming of man.

The first crude cells to populate the earth resembled bacteria. Bacteria evolved and combined with other early fragments of living matter to form the first plant cells.

Plant cells solved a serious problem which faced the living cell. About half the time, any given spot on earth was in darkness and there was no light energy available to sustain the living process.

However, the plant cell, through electromagnetic-chemical magic, captured and stored the fleeting rays of the sun as food. The plant cell could live upon this food during the hours of darkness. Plant cells were then, as now, the link between the sun and the whole of life. Without the wondrous plant cell, all life would disappear from the face of the earth as the images fade from a television screen when the switch is turned off.

Simple plant cells, the algae, multiplied throughout the early oceans, causing the waters to take on a noticeably green tint. From these single cells, plantlife evolved in dazzling complexity. Countless forms appeared, some staying in the oceans, but many finding their way across the continents which rose above the waters.

The efficient plant converted light energy into seeds and fruit and pollen, far in excess of that needed for its own survival. This made available a form of stored energy which could be a food for other forms of life. Plants in their living process consumed much of the early earth's carbon dioxide and released oxygen to the atmosphere.

The plant's production of food and of oxygen might have been a cosmic accident or could have been the pre-planning of a remarkable life system. But it was the presence of the abundant "stuff of life" produced by the plant world and the increasing presence of large amounts of oxygen in the atmosphere that were made to order for the coming of a new kingdom of creatures, the animals.

Animal cells evolved from plant cells in the microscopic world of one-celled creatures. The way this probably happened can be seen in a one-cell living "thing" existing today, which is both plant and animal. It is *Euglena gracilis,* or just euglena for short. Euglenae frequently make up the green scum that covers stagnant ponds. A euglena is a plant in that it contains chloroplasts, small manufacturing centers containing chlorophyll, which can produce food from sunlight. With its primitive eye, a euglena will swim over to a spot of maximum light in order to manufacture its own food. However, the euglena is also an animal, since in the absence of sunlight, it is capable of movement and can capture and ingest food particles.

It could have been in the slime of some primeval pond that groups of cells like the euglena moved entirely out of the plant kingdom. The change would have taken place at a time when generations of plant life had created an abundance of food matter in the water. It was then that our most ancient forebears, one-celled animals, freed themselves from the stationary existence of the plant kingdom, at the same time depending on plants for their food. The animals were to become an aggressive form of life, fighting for a limited supply of food, but they would have freedom of movement in exchange for a more uncertain existence.

The growth of animal life is often represented by a living tree. The tree of life is not like a pine tree in a forest where the lowest branches die off as the tree grows upward with time. The tree of animal life is more like a fruit tree whose branches twist and gnarl as the new foliage grows. Its branches thicken unevenly with the constant probing ever outward and upward toward the light. The tree of life on page 27 shows how animal life evolved and grew. This is a terribly sketchy way to express the billions of acts of creation, the billions of evolutionary experiments, and the living and dying of billions of creatures which brought our life system to its present state.

No book or volumes of books can do justice to the real story of how all the species of animal life came to be on this planet. All we can hope to do is to show that life moved purposefully toward more awareness and more mobility.

The first animals were the protozoa, and they inhabited the waters of the world 800 million years ago. Although consisting of only one

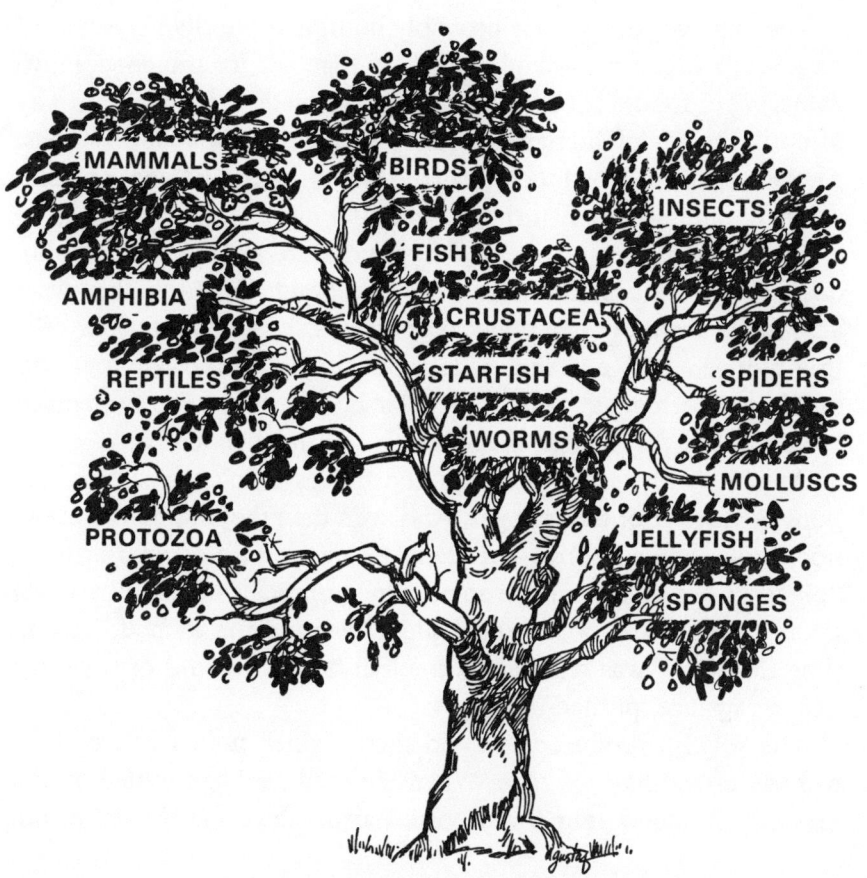

SIMPLIFIED TREE OF ANIMAL LIFE

cell, they had many traits of the larger animals that would follow. They could protect themselves from harm; they developed outer shells for protection; they could swim about by moving their hair-like appendages. As true animals, they captured, devoured and digested food particles in order to stay alive.

The first "organism" was probably nothing more than colonies of single cells which persistently clustered together for the cooperative gathering of food, for protection and reproduction, in much the way primitive tribes of humans cluster together for the same reasons. These "organisms" first took form in some warm, quiet pool of water where nothing would disturb their communing.

Then, since most waters of the world were not quiet and calm, cell colonies needed to form a structural bond. With this they were able to live beyond quiet pools. A sponge is such a colony. The cells in a sponge can live as individuals, but they prefer to live together, building a permanent "castle" of their bodies. Sea water is pumped in and out of the sponge to feed each cell independently.

Animals with specialized cells and tissues didn't appear for millions of years after the sponges. And even then they sat on the ocean floor looking like nothing but plants to a casual observer. But these "plants" were able to create some minor miracles of mobility when it came time to reproduce. The "plant" would form a "bud," and in time the "bud" was released as a jellyfish which could drift on the ocean's surface, pushed by the winds.

The jellyfish produced ova and sperm which united in the water to form a seed-like package. When the seed package settled on the ocean floor, a new animal was born, starting the cycle all over again.

The next step in nature's pursuit of mobility was the development of sea worms. They were the first to have a nervous system and a brain and are the ancestral creatures for all the higher animals of today. They represented a giant step in creation's plan because they could move around in a directed way. Yet, they were grotesque in appearance by humanity's standards. *A great many people would shudder with horror just looking at these squirming creatures, which testifies again to the lack of appreciation we, in general, have for the wonder of life.* Nature's body structure designs for all the most important animals in the world were tried out in the sea worms.

In order to survive in the buoyant ocean waters, a creature did not need to be weighed down with a bony skeleton. In fact, some of the animals best adapted to the ocean had no skeleton at all. Yet, in those early days in the ocean, nature seemed preoccupied with experimentation with all sorts of interior and exterior skeletal designs. Certainly there were additional survival advantages to creatures with a skeleton, especially if they were living near the ocean's shore where the crushing surf would damage tissues not protected by bone or shell. But that was only part of the reason for nature's probing.

A special, creative driving force pervaded all life in the ocean because of the presence of large continents where plant life had taken hold and flourished. Since all earth's life was interconnected, like one giant living cell, say our biologists, there was enormous pressure to bring animal life out of the ocean onto land. Thus, in order for the delicate, soft tissues of life to exist out of the sea, they needed to be supported by, or encased in, bone or shell.

Hundreds of millions of years' probing with hundreds of millions of creatures finally payed off. Somewhere around 400 million years ago, a crablike creature, probably some form of sea scorpion, crawled out of the sea. Complete with eyes and legs, this forerunner of the spiders and insects quickly adapted to life on land.

About this same time, life in the ocean had evolved the bony fish, with scales for outer protection of the body, and a highly developed inner skeleton which would support life's tissues on land. Some time after the shelled creatures had established themselves on land, the first fish ventured onto land, developing crude lungs and legs as they went. From the fish that found their way to land came the amphibians, the frogs and salamanders, and the reptiles.

There was little competition on earth's continents from other animals. Those first creatures out of the ocean, as awkward and clumsy as they were, had millions of uninterrupted years to evolve into animals that could survive comfortably on land. The direction of evolution, toward more awareness and more mobility and a higher possibility of survival, went steadily forward with these new animals on the earth's continents.

Since one of the major reasons for the evolution of a skeleton was to prepare life for an existence on land, there should have been an evolution of creatures adapted to ocean living which had no bony skeleton. And, of course, such creatures can be found in abundance in the ocean.

The molluscs are an enormous family of shelled animals. In their number are more species than among mammals; clams and oysters are among the best known. The molluscs usually live in shallows and along beaches, where their shells can protect them from the pounding waves and rocks and coral. The burden of their shells also serves as a protection from many predators.

However, the molluscs fulfilled their evolutionary urge for freedom of movement and awareness. Some of their kind abandoned their shells; their modern day survivors are the octopus and squid. These creatures, in addition to being swift of movement, have eyes that can see real images.

The squid and octopus are among the most successful creatures in the ocean. They came a long way from the average mollusc, a blind organism which could at best move only slowly across the ocean floor, carrying its burden of shell. Yet, if removed from the water and placed on land, a squid or an octopus becomes a soft, helpless blob of matter.

Also helpless, if removed from the water, is the king of the ocean jungle, the shark. Yet, the shark is not a fish in the true sense. The shark does not have a fully developed skeleton. It has only a backbone and enough cartilage to give it the proper rigidity in the water. The shark is a fearsome creature in the sea, but it has a body structure that could not adapt to land.

The first animals on land did not roam the earth; they crept and crawled about in a very humble way. Earth's animals, however, soon adapted to the ocean of air surrounding the earth, as yet another means to provide freedom of movement and insure survival.

The insects were the first creatures to fly. On the evolutionary scale of things, the speed at which crawling insects developed wings was quite rapid. *The development of thin, diaphanous wings to make flight possible is an example of life's powerful creative drive toward mobility. It is implausible to discuss the first change in an insect, which would lead to wings, on the basis of accidental genetic change.*

The reptiles, the first important vertebrate animals to adapt to land, evolved species with leathery wings for flight. The reptiles, particularly the dinosaurs, dominated the landscape for 200 million years. However, the reptiles had inherited from their ancestors, the fish, a cold-blooded metabolism, which severely limited their means to adjust to climate. But with life's fantastic diversity of creatures and with millions of years for experimentation, improvements in the animals could be expected.

The problem of cold-bloodedness and lack of mobility in the reptiles was largely overcome by the evolution of birds. A bird's body clearly shows that it evolved from reptiles, but the evolution of a bird from a reptile cannot be explained by genetic accident without insulting the intelligence of the reader. Too many carefully orchestrated, inexplainably complex changes had to occur to evolve a bird from the form of a reptile. Even the passage of tens of millions of years could not allow for birds to evolve by accident. The development of a warm-blooded creature with hollow, light bones, and feathers, and wing structure, had to proceed purposefully if a bird was to be. There were many design failures, and whole species perished in the process. Yet, the ceaseless probing for mobility and warm-bloodedness continued until birds came into existence 150 million years ago.

The mammals also evolved from reptiles. Like birds, they developed warm-blooded bodies, but instead of feathers, they developed furry coats for temperature protection. This gave the mammals an ability to range over all the earth in every climate. The mammals became such efficient animals that, millions of years ago, they crowded most of the reptile species to extinction.

The ceaseless probing to create new life had gone on for billions of years. Thus, millions of years before any sign of man appeared, the tree of life had come to full flower. It was like an apple tree in the spring. Its branches showed thousands of blossoms, each flower representing a carefully perfected species of life. Each form of life was its own miracle of creation and had found its own special place and function on this planet. Blind energy and matter had become plants and animals of infinite variety.

Plants had adapted to every condition of earth's climate. Microscopic plant cells floated on sunlit surface waters of the ocean currents, in fresh water ponds, and in streams and rivers. Plants of all sizes and descriptions also lived in the shallows of every fresh and salt water body of the world. Plantlife, from mosses to trees, was equally abundant on the grand continents which had risen from the ocean. Plants thrived in hot humid jungles, in hot dry deserts, in the biting cold of the arctic regions, and in the wind and cold of the mountain tops.

Protozoa still inhabited the earth by the billions. Tens of thousands of species of worms pursued a quiet existence on earth, living reminders of these early forms which served as a necessary intermediary to all the higher forms of animal life.

At least 900,000 species of arthropods still lived, ranging from the clams and lobsters of the sea to the spiders and insects which could be found everywhere.

Living fossils almost a billion years old still flourished in the oceans—the jellyfish, corals, sponges and starfish. The molluscs, which had been living in the waters of earth for hundreds of millions of years, were now represented by fifty thousand separate species.

Over ten thousand species of fish of every size, color and adaptation were represented. Some had complex behavioral patterns and remarkable means for navigating.

Thousands of species of amphibians lived a dual life on land and in the water, living comfortably in éither. They filled the shores of every pond and lake with their remarkable life style. Though no longer masters of the earth, six thousand species of reptiles still were living on the lands or in the waters as remnants of a dynasty that had once dominated the earth.

Several thousand species of mammals inhabited the lands of earth. The adaptable mammals had found ways to live in the hottest and the coldest parts of earth. Whales, walruses, seals and porpoises had returned to the sea, and the bats had even mastered flight.

Thousands of species of birds graced the air, water and land. In these highly developed creatures the spirit of life had found the highest meaning. The billions of years of constant probing to produce the miracle of vision with color, depth and accuracy was no more vividly expressed than in the birds. The larger birds had a power of vision that would never be matched by any other animal. The birds' power of flight provided their superior senses with a grandeur we can only imagine. Nowhere was the spirit of life to burn more brightly with meaning. The bird was a life form which, in its everyday existence, could savor everything this planet had to offer.

Imagine the kaleidoscope of life's richest impressions which are a natural part of a duck's life. The duck can move with splendid grace on the water or in the air. It can dive under water with the speed and mobility of a fish. It can walk ably, though awkwardly, upon the land. It can mingle with the richness of life on a pond sparkling in sunlight or it can soar far above the waters and the land to gain a perspective on the vistas below. A duck can choose at will any corner of the earth to rest, to eat, or to relish the experience of a new location. The duck, with unerring accuracy, can navigate thousands of miles to enjoy the best of the different climates of earth. The flame of energy which found meaning in life could be content for an eternity in these creatures alone.

Nature had achieved a harmony millions of years before any creature resembling man had appeared. Life had become one with the waters, lands, and winds of this special planet. The experiment of life had been a success, and all of life's representatives were casting about with untold billions of pollen grains and spores and seeds and sperm and eggs in the fierce desire to turn every ray of sunshine into a plant or an animal.

Life on earth had achieved perfection without mankind and civilization. Yet, in nature's harmony there was a waste of life's potential; life was limited to one planet. Millions of creatures were perishing daily for lack of space. The living earth was like a cottonwood tree which, every summer, launches thousands of seeds into the air; but because of competition for space and light, perhaps only one or two seeds find places to grow.

Meanwhile, in the night sky, billions of stars, other suns whose planets were barren, quietly beckoned with their life-giving light.

Chapter 4

A CREATURE OF PROMISE

> When I consider thy heavens, the work of thy fingers,
> the moon and the stars, which thou hast ordained;
> What is man, that thou art mindful of him?
>
> Psalm 8

A ground-hugging plant or a towering tree spends a lifetime confined to the spot of soil where it took root. Yet, when it is time to prepare its seeds for journeys to new places, a plant seems to know the ways of the waters and the winds. The dandelion launches its seeds attached to perfectly designed and incredibly efficient air foils. The palm tree's seed, the coconut, will float and drift for thousands of miles on the ocean, find a new shore, and renew life.

A dandelion plant could not experience flight. A palm tree anchored to one island could not experience ocean currents or visit other island shores. The dandelion and the coconut palm are a part of the wisdom of the life system whose creative powers reach across all species of life and know of the earth and the universe.

If nature can empower the winds to spread the seeds of a stationary plant, then all earth's resources should be able to find a way to disperse life's seeds to the universe.

In the natural, harmonious world of ten million years ago, nature had no way to carry life to another planet. A new kingdom—a fourth kingdom—had to evolve if life were not to be trapped on earth. The creatures of the fourth kingdom had to consist of stuff unlike living tissue, yet be lifelike in form and function. Creatures of the fourth kingdom would need to develop great power for movement and travel. Somewhere in nature's profusion would be a creature or class of creatures which could evolve to form a fourth kingdom.

As spectacular and as varied as the birds were, none of their endless numbers had the starting characteristics needed to depart from a natural existence. The birds were creatures which could take full advantage of the air, the waters, and the land of this planet. In their beautiful adaptation, birds could be fulfilled in earth's nature only.

We know now that the fourth kingdom was to emerge from the mammals, consisting of many varieties of animals that had adapted to suit almost all nature's purposes. The fourth kingdom started in the lush tropical forests of the earth among the primates, which included the monkeys and the apes.

For over ten million years, the primates had found an idyllic existence among the high branches of trees. The primates did not need four-legged speed to obtain food or to avoid being eaten as did the ground-living mammals. They had developed elongated fingers and toes for grasping the limbs and vines when living in and traveling through the tree tops. Skilled hands were necessary since a fall could be fatal. Fingers became more agile as they were used for grasping food and caring for the young among the high branches.

Swinging through the trees required keen vision and good depth perception. The need to find colorful fruit for food among the dense foliage caused the primates to develop color vision, common among the birds, but uncommon among the mammals.

High intellect, manual dexterity, and keen all-purpose vision made the primates logical candidates to carry out creation's plan. The pull of the stars and the push of life to survive would cause some dramatic changes to begin among the primate species.

Modern day paleoanthropologists have not produced any convincing reasons why some primates left the safety and abundant food supply of the trees to walk on two legs. A blind force of survival would have favored primates which stayed in the trees. A few claim apes were forced to the ground by a worldwide reduction of trees.

If life on the ground became a necessity, the primate types most likely to survive would have been baboon-like creatures, capable of fast, four-legged locomotion with powerful jaws to survive in competition with the other ground-dwelling animals.

Rather, there suddenly appeared, between five and ten million years ago, two-legged primates that walked upright with an awkward gait and possessed a brain larger than that of any other ape. They were pre-men, and their origin appears to have been in Africa. The appearance of a number of species of two-legged animals seems to have occurred, revealing the normal experimentation by nature, where varieties of similar creatures are tried in order to reach an evolutionary goal.

Pre-men appeared in great numbers during some of the greatest climatic upheavals earth's life had ever experienced. The Pliocene epoch with its droughts was followed by the ice ages of recent geological time. Climatic changes were especially severe in the last million years which heralded the coming of modern man. These enormous swings in climate have been attributed to variations in the sun's radiation by most of today's scientific community. It was during these irregularities in the sun that pre-man and then man appeared on earth. *The life system, a product of the sun's energy, could sense trouble from its star, and would prepare a means of escape.*

We assume that our larger brain gives us intelligence superior to all other animals. However, animals have better mental powers than we for survival in the natural world. Many animals have sharper vision, better hearing, better developed senses of smell and touch, and better muscular coordination than man. Birds have what appears to be supernatural abilities for long distance navigation. Some animals have orders of behavior so much above our own mental processes that we cannot understand them. Unique mental abilities in other animals are often set aside as "instincts" to preserve for ourselves an inflated and distorted version of human intelligence in the animal world.

On reflection, we must admit that our sense and our animal in-

stincts are those of average mammals, which allow us to just get by in the animal kingdom. The additions to our brain, which set us aside from the other primates, evolved in order to carry a special burden. The human brain was required to create and nurture the fourth kingdom, creatures of metal, stone, and plastic. The fourth kingdom could not evolve directly from organic life. The human primate became the custodian of the required brain cells in order to be the custodian of the fourth kingdom.

Over two million years ago, we could have identified a creature of promise emerging. The walking pre-men of Africa had begun to free themselves from the balance of nature. They had begun to create and care for the first species of the new kingdom. They had started to fashion weapons or tools from stone or wood. Rocks held in the hand were roughly shaped; sticks were pointed; fabric and rope were made by weaving vines and foliage. These simple items started our new creature on his way. Rocks and sticks served to assist in the protection of this two-legged animal, who was not going to survive by brute strength or speed.

A cornerstone of human behavior was set down at our very beginning: *Man would rely on the fourth kingdom for his very survival.* Tools and weapons would be so important that they would be improved as human talents improved. As man realized his dependence on these new creations, he would carry them from generation to generation as an intimate part of his life. Men had compelling reasons to use their larger brains and to increase the manipulating skills of their hands.

As the human primates spread over the earth, they were subjected to a whole range of environments, especially to extremes in climate. It is a zoological fact that animals subjected to such a wide range of conditions break up into different species in order to survive in their environment and to function more efficiently. Yet, strangely, as this new animal approached the form of modern man, he remained as one species, capable of interbreeding across all the earth. The zoologist would call this an indefinite boundary of interfecundation, an ability to interbreed readily across a variety of races.

The evolution of mankind as a single species demonstrated a reversal or a stoppage of the normal evolutionary process. It was apparent humanity was not to be a new form of life which was destined to adapt to the nature of earth. Rather, this departure from normal evolutionary law indicated a creature that would depart from nature's usual course. Humanity was obviously being prepared for a global role of great significance.

Man showed himself again to be the creature of promise when he found the power of fire. A warming fire at the mouth of a cave was the first feeble step in controlling a release of energy far greater than could be generated by living muscle. The sun's energy, stored in the bodies of plants or animals, was accumulated over a period of years or centuries. Our new animal in even his early stages could release this stored energy in minutes through his newly-found control of fire.

Man soon learned to use fire in order to survive. Fire kept him from freezing and made his food more edible. Thus over hundreds of thousands of years, the humanoid brain became deeply imprinted with the importance of the controlled release of energy.

These important ingredients for space travel, artifacts and fire power, were to stay forever in this new animal's life because they had become the essence of his survival. Man left forever his primate Garden of Eden, left the safety and the plentiful food of the trees. He was entrapped in a culture turning rapidly away from earth's nature, and he would become more ingenious with his artifacts and with sources of energy in order to survive.

The life system had knighted this animal, and the talents that would some day create the fourth kingdom gave this primate great power. The endowment of one species with such power would give one animal reign over all other creatures on earth. This could be a hazardous experiment by nature. The endowed primate might not recognize the purpose of the power given his species, nor use it properly, and nearly all nature could be destroyed. Yet, the needs of the life system with its flickering sun were critical. The attraction of lifeless stars and planets was strong. Thus, the probing by creation would proceed onward.

Greater things were expected of this two-legged primate than producing simple artifacts and cave fires. Thus, an improved creature with an enormous brain evolved about 100,000 years ago. The form of modern man had appeared. *Homo sapiens* at last! The mental capacity to support the fourth kingdom had been put in place.

From 100,000 to 35,000 years ago, the first simple species of the fourth kingdom grew in profusion, like the first simple forms of life in the ancient oceans. The tools and weapons, though simple, were now in many more forms of stone, wood and bone. Axes, borers, choppers, knives, scrapers, saw-like blades, chisels and planes were the "protozoa" of the fourth kingdom.

Strong legs allowed man to walk great distances, and adaptation to a hunting life allowed him to spread over all the earth. Man existed about 35,000 years ago in small tribes and most humans then were imprinted with the same pack instincts as wolves and baboons. Unfortunately, great numbers of other animals, whole species, perished under the hunter's onslaught so that humanity might survive.

A new skill was added to the talents of man between 35,000 and 10,000 years ago. Cave dwellers found a way to convert mental images to art forms through eye-hand coordination. Humans began to draw objects and make small statues of ivory or bone. Archeologists have found countless examples of wall paintings and statuettes which prove the ability of our cave dwelling ancestors to convert thoughts to art forms. The ability to convert mental images to art forms was a critical new talent. These skills would be needed in a society which could build complex machines.

Cold winters existed over much of the earth during the last glacial period. However, there was an abundance of animals, and, as organized hunters, men easily found a ready source of food. Also, our ancestors had plenty of time in their fire-warmed caves to develop cultural skills. For thousands of cold years, people sat around cave fires, like men around the stove at a country store or women at a sewing circle, and developed a spoken language.

As a predatory hunting animal, man's territorial instincts were especially strong. The number of people in a given territory was limited to avoid depletion of the food supply. Like other hunting animals, primitive humans fought ferociously to hold territory against invasion. Man appeared destined to remain a society of small, scattered tribes.

Great changes would be needed for humanity to carry life beyond the earth. People would need to work in an immense, densely packed, and closely coordinated society. Although *Homo sapiens* was a creature of great promise, the way to this society was unimaginable 25,000 years ago.

Chapter 5

EVOLUTION OF THE FOURTH KINGDOM

> If one were to ascribe a single objective to evolution, it would be the perpetuation of life. The entire strategy of evolution is focused on that single end.
>
> George M. Woodwell,
> "The Energy Cycle of the Biosphere" [1]

Civilization started when mankind found the way to obtain food by farming instead of by hunting. The effect was to free people from the constant search for food, a problem faced by all of nature's animals. So, more humans had time to be creative with their enlarged brains. There was more time for reflection, more time for work at the mechanical arts, more time for developing a written language and probing into mathematics. Farming societies that did not have to roam about in search of food, began building permanent shelters. Villages emerged about ten thousand years ago in many parts of the world.

Territorial tensions between villages were high. These early villagers were people whose ancestors had spent hundreds of thousands of years as animals close to nature. The residual instincts of hunting animals caused the villagers to distrust strangers and to fight ferociously if there were a threat of invasion of territory.

---
[1] Woodwell, George M., "The Energy Cycle of the Biosphere," *Scientific American*, September, 1970 (c. 1970, Scientific American, Inc.), p. 64.

Yet, millions of people had to live and work together if machine technology were to evolve.

As the farming villages grew and prospered, a strange behavior erupted. Leaders arose with a blood-thirsty ferocity to conquer more territory and more people. Homes were burned and plundered, sacrificed in battles of conquest. People who fought to defend their territory were ruthlessly killed. This newcomer to the animal world was killing many of his own species. Man seemed an animal gone berserk.

The history of the early agricultural civilizations of the Mideast is reflected in the Old Testament of the *Bible*. Conflict and conquest were common themes. Some writers of that time concluded that the destruction and bloodshed were God's will and God's way. Some believed God was being wrathful and punishing the people for their offenses.

These faithful people recognized it was the hand of creation at work, but they didn't understand the reasons. The people of the Old Testament did not recognize war was a means to build powerful nations where millions of people could be brought together to build large, complex machines. They did not visualize the supernation that would come into being regardless of the conquest or ruin of countless tribes, villages, kingdoms and nations. Territorial barriers and cultures would be obliterated in the process.

Through war and conquest, cities and nations evolved as early as six thousand years ago. The city cultures took an intensely religious and mystical interest in the heavenly bodies beyond the earth. One such city was Ur, birthplace of Abraham, the first patriarch of three major religions—Judaism, Christianity and Islam. Ur was located in the fertile crescent of the Mideast in the lowlands of the Tigris and Euphrates Rivers and was best known as a center for worship of the moon. Other cities of Ur's time contained temples dedicated to the sun, stars or constellations of stars or planets. *The citified cultures of man were beginning to perceive their destiny.*

The size of man's battles grew larger as the building of nations grew to the building of empires—the Persian Empire, the conquests by Egypt and Greece, and the Roman Empire. *The empires were nature's probing to build a supernation so bountiful with people and resources that any technological feat would be possible.*

The new empires created great cities: Athens, Rome, Alexandria, Syracuse, to name just a few. In these great cities, a new breed of humans became comfortable living apart from nature. People became dependent on the cooperation of their fellow man and on the emerging technology.

The cities contained societies which would collapse if their technology failed. *Over three thousand years ago, the cities carried on experiments in the trust in technology which would be needed by colonies of people who would some day set off toward a distant star.* Central water systems and multifamily dwellings were commonplace. Travel by foot was supplemental to travel by chariots and boats. Some cities developed elaborate fortifications for the common protection of their people.

As war and conquest became commonplace, and survival became uncertain, man, by instinct, worked at his artificial creations with even greater urgency. As the empires established themselves, the fourth kingdom multiplied at an enormous rate. The wheel was discovered. A way was found to mine and extract metal from ore. Copper, bronze, gold, and silver were shaped to countless uses. Large and varied buildings came into being. Boats and ships of all sizes and shapes were built, which could move by the power of muscle or the winds.

Only a few people who lived twenty-three centuries ago are remembered today. But one man in particular illumines the history of the third century B.C. He was Archimedes, from the Greek city of Syracuse in Sicily. Archimedes' talents were in harmony with the destiny of man; he was both a mechanical genius and an astronomer.

Archimedes described the laws of mechanics explaining levers, pulleys, gears and hydraulics. He invented advanced machines of war, which at times almost terrified the Roman forces laying siege to Syracuse. However, Archimedes, the son of a famous astronomer, took more pride in his studies of astronomy than in his mechanical ability. In his book, *On Sphere Making,* Archimedes showed great insight regarding the motion of the sun and the planets.

Man's interests in mechanics, science, and astronomy as exemplified by the life of Archimedes were to be repeated many times in the history of civilization. Astronomy was the first organized science, and from astronomy came not only the principles of mechanics and mathematics, but also the concepts and measurements of time and distance. *The astronomers, the men who looked out at the stars, were also a wellspring of ideas which would lead the way for life to reach the stars.*

The closest the earth had come to a supernation two thousand years ago was the Roman Empire. Although fashioned from conquest, the Roman Empire spread technology and culture to the dark corners of Europe. The Roman Empire communicated civilization's latest achievements to many persons.

While the Roman Empire was in full flower, a man who was to be known by his followers as Jesus Christ was born, lived and died in one small corner of the domain. His philosophy and teachings were critical to creation's plan. Jesus reemphasized the earlier codes of behavior, especially the Ten Commandments, which allowed people to live together in peace. However, his teachings stressed love of one human for another. This was an exceedingly powerful way to subdue the savage territorial hunting instincts residing in the human brain. Jesus proclaimed that all who believed in Him were like one body. This philosophy would allow thousands, and then millions, to work in harmony for a given cause.

Christianity barely survived during the first few hundred years of its existence. It found its way to Rome, the heart of the Roman Empire where Apollo, god of the sun, and Diana, goddess of the moon, reigned. The Christian religion slowly grew in Roman society, but with a high loss of lives.

Then a miraculous, well-documented event occurred. The Roman Emperor Constantine saw a cross superimposed over the sun before the beginning of a decisive battle. After Constantine won this battle, his conversion to Christianity was rapid. And it was followed by Christianization of the Roman Empire.

Christianity had come to Rome as a variation of Judaism, which emphasized the supremacy of man, a divine destiny for man, a belief in a personal God, and love of fellow men. However, Christianity became Romanized in the process of being accepted throughout the Empire. Many of the original traditions were radically altered. Judaism was blended with the beliefs in the Roman, Greek, and North European gods representing the power of nature. For example, the Roman Christian day of worship became "Sunday" in memory of the power of the sun and the god Apollo.

It appears that the Roman form of Christianity, based on a unique blend of ideas, was one reason for the rapid rise in science and technology among the Christian societies of the West. This new brotherhood allowed great numbers of people to live and work in social harmony, while at the same time they recognized the wonders and the power of nature which would encourage scientific pursuit.

The collapse of the Roman Empire in the fourth century caused a pause in the advance of civilization, technology, and the fourth kingdom. However, the Roman Empire had scattered ideas and concepts over all Europe, like a withered, dying plant casting off its seeds on a cold, dark and windy day in fall. During the centuries of cultural darkness that followed, the ideas remained alive. New nations grew from the ruins of Rome to become strong, organized societies of thousands of people who would again take up the arts and sciences. By about the tenth century, the northern Europeans were making steady contributions to the life of the fourth kingdom.

EVOLUTION OF THE FOURTH KINGDOM 47

From the eleventh to the fifteenth centuries, the rate of evolution of the fourth kingdom increased dramatically.

Until men began to respond to their biological destiny, there was no earthly reason to measure and record the passage of time. All life was in tune with the natural rhythms of the days and nights, the tides, and the seasons. The plants and animals had well developed internal "clocks" to pace their living habits.

Then came the astronomers of the Middle Ages who found that in order to measure the movement and predict the positions of the heavenly bodies, they needed to measure time precisely. The first mechanical clock was built in 1088 by the Chinese Mandarin, Su Sung. This clock was also a planetarium which displayed the motions of the heavenly bodies. Europe's first mechanical clock, built in 1362 by Giovanni de Dondi, also recorded the movements of the planets by a complicated set of gears.

The physical sciences which followed astronomy were possible because of the invention of a precise clock. Clocks then became a means to synchronize the efforts of millions of people. The advance of culture and technology rapidly increased in time with the tick of this new device, the clock.

Yet, very few of us remember that the original reason for the clock was the science of astronomy, the looking outward to other planets. *And even fewer of us see in the face of a clock mankind's purpose in this world.*

The astronomers emerged in the fifteenth and sixteenth centuries with the aid of the telescope, which gave them for the first time a look at reality in the universe about them. And they ran afoul of the conservative leaders of the early Christian Church. The Polish astronomer, Copernicus, ironically a member of the church hierarchy, announced one day that the earth was not the center of the universe, but rather that the earth orbited the sun. An immediate uproar was heard from among the leading cardinals and bishops of Europe. Then Galileo, an Italian astronomer, turned a telescope to the skies in 1609 and concluded Copernicus was correct. For this, Galileo was placed under permanent house arrest.

Nothing gives more publicity to a position than the violent opposition of the establishment, and the church's opposition brought astronomy into the limelight. The passage of time proved the astronomers correct, and these studiers of the stars and nature took a place of prominence in the culture of Europe.

The mysteries of the human spirit, the complexities of civilization and the workings of creation are best understood when they can be described in the personality and life of a person. The man or woman may be real or mythical. The ancients understood the power of personification well. The earliest civilizations knew the sun to be, somehow, the source of life and thus the source of much power and good. The ancient Greeks and Romans personified all this mystery in the form of a man-god, Apollo, in order to allow such abstract ideas to be accepted by great numbers of people.

Modern societies continue to symbolize the living power of creation, often called God, in male human form. One remarkable man was Yeshua of Nazareth. He embodied so much of the essence of God in his life that he became known from the Greek and Roman translations as Jesus Christ. He personified God's presence in humanity to his followers and to the Christian Church.

The workings of creation were also personified in the incredibly illuminating life of Leonardo da Vinci during the fifteenth century. *Leonardo's talents, ideas and accomplishments all point to a creative force, working through the brainpower of humanity to embrace the fourth kingdom.*

Leonardo, in his early career, was a superb artist and sculptor. His painting of the *Last Supper* was sufficient to give him lasting fame. He totally embodied this great creative passion of humanity—art. Leonardo thus represented man's ability to describe a thought with a painting or a sculpture, which was an enormous step forward for one of nature's creatures. Since the days of the cave man, humanity has been celebrating the existence of art with a delight and an obsession without knowing exactly why.

As Leonardo passed the age of thirty, he set his art aside more and more as his interest in mathematics, science, engineering, astonomy and architecture grew. Leonardo had evolved from an emphasis on art to an emphasis on science, yet he always embraced both, as human civilization would continue to do centuries later.

Art was the beginning of modern technology. The fourth kingdom emerged when paintings evolved to architectural and engineering drawings. The fourth kingdom became possible when crude symbols became numbers, alphabets, mathematics and the printed word.

Leonardo despised the "bestiality of war," yet some of his most ingenious inventions were war machines. Leonardo worked for the Duke of Milan and Cesare Borgia in their battles of conquest. Leonardo thus represented all human societies, before and after his time, that suffered from war, but were stimulated to great achievements by the experience.

By Leonardo's time, the first simple machines had evolved. Cams, levers, gears, and wheels had been joined together to drive a battering ram, hoist a heavy stone, or keep time. *These early machines evolved like the animals in the ancient oceans. Cams, levers, gears, and wheels were like the single celled animals that clustered together to form a sponge or a jellyfish. Machines would be expected to parallel animal evolution because they were the product of the accumulated brainpower of the animal kingdom.*

Machines, to fulfill their destiny, required a power of movement of their own, independent of the wind or muscle power. Two hundred years before Leonardo, the Chinese had used rockets and gunpowder —the first known use of a heat engine—in their battles against the

Mongols. Shortly thereafter, a second heat engine—the cannon—was introduced in Europe. The cannon both popularized the art of cutting metals, and destroyed a feudal system which had fragmented nations and frustrated all attempts to create larger and more powerful nations in Europe. *The cannon was a manifestation of the striving of nature to find metallic machines to produce power and create greater organizations of people.*

Leonardo was well versed in the science of his day, and his thinking mirrored nature's plan. Leonardo designed over eight hundred machines, many of them predicting the day when they would run on their own self-generated energy. Leonardo foresaw the automobile, the submarine, and even the airplane. *Leonardo's talent was a sign that the human mind was being prepared for the day when coal and oil would usher in the Industrial Revolution. The evolution going on in Leonardo's mind was like the evolution of the animals in the ocean which developed special body structures in preparation for the day they would walk upon the land.*

The first powered machines of significance were coal-fired steam engines which appeared about 1700. Over the next two hundred years, steam power was adapted to locomotives and boats, followed by crude attempts at powered flight. Powerful machines began to appear in great profusion. The emerging Industrial Revolution paralleled Teilhard's statements about organic evolution: "This groping strangely combines the blind fantasy of large numbers with the precise orientation of a specific target," and "pervading everything to try everything and trying everything so as to find everything." [2]

History looks upon the Industrial Revolution as the beginning of material riches and leisure for mankind. *However, the pulse of nature*

---

[2] Teilhard de Chardin, Pierre, *The Phenomenon of Man*, (New York, Harper and Row, 1965; c. 1959 William Collins Sons & Co., London, and Harper & Row, Publishers Inc., New York), p. 110.

beat faster for quite different reasons. Authors and poets, as if on cue, began to write about travel to other planets. Poems about space travel appeared in the mid-1800's, and Jules Verne in 1865 wrote about a fictional trip from earth to the moon.

The Industrial Revolution placed the societies of Europe and America in a totally new relationship with nature. Machines and technology provided better shelter from the elements, more efficient and reliable food production, and better protection from man's natural enemies. The population of the advanced civilizations rose dramatically. Many people had entered a new situation where most of their society would perish if machines and technology perished. Man and machine had entered into a symbiotic relationship, a phenomenon found commonly throughout the life system.

Symbiosis is the insoluble partnership of unlike organisms which results in the betterment, and more often, the very survival of both. Certain fungi are cultivated by beetles, termites, and ants for food. Animals cannot survive unless bacteria and fungi live intimately within their bodies, often within their very cells and tissues. The symbiotic bond between bees and flowering plants is known to every school child. Man relies on domesticated plants and animals, and as a hunter he formed a symbiotic bond with dogs tens of thousands of years ago.

A most remarkable symbiosis in nature is found in the lichens. The lichen was originally considered a plant similar to the mosses. But the lichen was unusual; it could live on the surface of barren rocks in extremes of climate. However, under the scrutiny of a microscope, the lichen turned out not to be a plant. The lichen was shown to be two unlike organisms joined together to create a form of life able to flourish in places which could not otherwise support life.

One symbiont forming the lichen is a fungus. The other partner is an alga. The fungus captures water from the air. The alga manufactures food from sunlight. The two exchange water and food and other substances to survive.

In the lichen, nature has intertwined the lives of two separate creatures to assure that life would occur in otherwise uninhabitable places. Similarly, the symbiosis of the human animal and creatures of the fourth kingdom also allows life to survive in unlikely places—someday in transit between the stars.

As creatures of our thought, machines could easily become extinct simply because of our apathy. Nature, however, reached out and provided the fourth kingdom with the protection of a symbiotic bond where the loss of one partner would mean the loss of the other. Most Americans today would die if they abandoned technology. Our nation as we know it would be destroyed. If the fourth kingdom were forsaken, tens of millions would die of starvation, disease or exposure to the elements. Therefore, the survival of the fourth kingdom means our survival.

The Indian people of North America controlled, during the early 1800's, most of the lands that were to become the United States. The Indian was in harmony with nature. A lack of technology required that the Indian people submit to the forces of the natural world. A tribal and territorial culture kept their population in balance with the rest of nature.

The Indian nations, however, were doomed by the newly formed United States, whose people and technology had been transplanted from Europe. As the United States grew stronger, it expanded westward. The spread of the European industrial civilization across this continent and its cruel effects on the Indian nations are poignantly recorded in the book, *Bury My Heart at Wounded Knee*. In this book, Dee Brown detailed the heartbreaking sacrifice of the Indian's culture, his leaders, his religion and his nations so that a supernation might take form. Mr. Geoffrey Wolff of *Newsweek* magazine said in his review of this book:

> The books I review, week upon week, report the destruction of the land or the air; they detail the perversion of justice; they

reveal national stupidities. None of them—not one—has shamed me as this book has. Because the experience of reading it has made me realize for once and for all that we really don't know who we are, or where we came from, or what we have done, or why.[3]

Mr. Wolff and other observers of the human scene can easily conclude on the face of it that our technological civilizations are misguided as we build our future on a pad of concrete, metal, glass, and plastic. A faithful conservationist may well wonder how God could allow a marauding technology to be the victor over a people so much in tune with nature. The Indian tribes, as did the Hebrew tribes of the Old Testament, had moments when they despaired of their God.

It was the same old problem since civilization began. There was no room for a tribal, territorial culture where a great industrial nation was needed someday to launch life from this earth. It was superior technology that destroyed the Indian nations. Countless other small tribal nations had been destroyed throughout man's history, and for the same reasons.

The nineteenth century politicians justified the ruthless conquests by the United States from one coast to the other under the doctrine of manifest destiny. They said it was God's will that a continental nation should be built. And indeed it was in accord with creation's plan, but not for the reasons envisioned by the early politicians, who simply wanted the power and wealth of a great nation.

If there is any doubt about a supernatural bond that held the United States together, the Civil War should dispel that doubt. The United States survived, but at a monumental sacrifice in lives. The nation's leader at the time, Abraham Lincoln, was a simple rural man, but he rose to incredible heights of leadership in order that the United States might remain undivided.

The expansion of Russia to a supernation, the U.S.S.R., has its own tumultuous "manifest destiny" story. Territories and people were conquered in order to combine material and human resources. *The*

---
[3] Wolff, Geoffrey, "Prairie Fire," *Newsweek*, February 1, 1971, vol. 77, p. 69.

goal again was an industrialized nation worthy of carrying the fourth kingdom forward.

A continuing onrush of science and technology ushered in the beginnings of this, the twentieth century. The more advanced civilizations by the year 1900 had gained a dominant position over all other forms of life. Civilization behaved as though the earth and its life were at its disposal, to be used as necessary to benefit humanity. Mankind thought itself apart and above all the rest of the natural world.

This arrogance of power in Western civilization was opposed to one man's philosophy that had survived for nearly eight hundred years. The sacred value of all life was basic to the teachings of Francis of Assisi, who lived in the years 1181–1226. He was founder of the Franciscans and became known as St. Francis to the Christian Church. Francis considered all nature to be the mirror of God, and the parts of nature to be so many steps to God. He saw all life as one family of creation. Francis taught that we have no right to be contemptible to or unappreciative of others of life's creatures simply because they are powerless.

It was ironic that the power of the fourth kingdom, which had been given to humanity in order to serve the needs of all life, was being used by Western civilization to dominate or enslave nature's creatures.

The fourth kingdom grew rapidly, due to the urgent demands put on Western civilization by World War I. By the 1920's machines could travel on land, on and under the sea and through the air at speeds unknown in nature. Authors began to remind civilization of its destiny. The comic strip adventures of Flash Gordon and Buck Rogers, visiting new planets in giant rocketships, caught the attention of millions.

Rockets were not new to civilization. They had been used by the Chinese in the thirteenth century. The national anthem of the United States recalls that "the rockets' red glare" illuminated the flag at Fort Sumter, South Carolina, during the war of 1812 against England. During the 1890's Herman Ganswindt, a German inventor, and Konstantin Tsiolkovski, a Russian mathematics teacher, were already dreamily talking of traveling to other planets by rocketship. Robert Goddard of the United States envisioned in 1918 a space ark which would take people to another star.

The new technology of the 1920's caused civilization to take rocketry and space travel seriously. Until the radio, there was no way to communicate with or control fast moving machines rushing through the atmosphere or through outer space. With the advent of electrical control systems, machines finally had a primitive "nervous system" and thus became even more lifelike, as well as more reliable and controllable.

Rocket and space societies had formed in Germany, Austria, the U.S.S.R., The United States and Great Britain by 1933. Scientists, teachers, and mathematicians, in addition to astronomers over all the world, began exciting experiments pointing toward interplanetary travel.

The scientific revolution began to slow down as the excitement of World War I faded away in the 1920's. It would seem another great war was needed to push man and his machines to develop a means to reach out to new planets.

And there arose from Germany a man, Adolf Hitler, with a savage consuming desire to conquer the world. While Hitler was at times irrational and evil in his personal motives, he served as a charismatic leader who plunged the civilized world into a great war.

People wonder today, both inside and outside Germany, how Hitler derived so much power. There is, of course, an answer that has been historically proven throughout human history. The explanation lies in the fact that Germany in the 1930's had the most advanced machine technology on earth. Even more importantly, Germany had the beginning of the capability to travel interplanetary space. As has been shown throughout man's history, a technologically superior nation will reach out to conquer as many other peoples and nations as possible to create an even greater nation. Hitler was simply a catalyst, unfortunately a very poor one, who unleashed the residual instincts in a people, forceful instincts that had existed in mankind since the beginning of civilization.

The great technological achievements wrought by World War II would fill volumes. Man had been molded to develop his machines with his most passionate creativity when survival was at stake. This was a pattern of response that was as instinctive in humanity as the migratory instincts are in birds at the change of seasons.

The emotions of this great war helped man to recognize that his machines were an important life form. The symbiotic relationship between man and machine finally came fully into human consciousness. During World War II, men gave their airplanes and their trucks names, and painted them with eyes and teeth to resemble animals. The best known American vehicle, the Jeep, was named after an imaginary animal of the comic strip world.

Perhaps the most widely read and respected writer of World War II was Ernie Pyle. Critics have felt that his writing on the return of an American bomber and its crew to an air base in North Africa, after being presumed lost, was and is among the most beautiful prose ever written. The bond of love that had developed between man and the kingdom of machines was never more sincerely expressed than in the few words used by Ernie Pyle to describe the event:

> Then we saw the plane—just a tiny black speck. It seemed almost on the ground, it was so low, and in the first glance we could sense that it was barely moving, barely staying in the air. Crippled and alone, two hours behind all the rest, it was dragging itself home.

EVOLUTION OF THE FOURTH KINGDOM 57

I was a layman, and no longer of the fraternity that flies, but I could feel. And at that moment I felt something close to human love for that faithful, battered machine, that far dark speck struggling toward us with such pathetic slowness.[4]

Ernie Pyle saw this airplane and its ten men as a single organism, a single identity. He captured the feelings of all who have had to entrust their lives to machines. *The bond between man and the fourth kingdom would need to be at least as strong as human love if we are to entrust to machines our future in the uncertainties of outer space.*

Remembering the millions who fought, suffered, or died in World War II, we may well wonder how this war could be credited with assisting nature's needs. Certainly the thousands of Americans who died in agony on the invasion beaches of Normandy in 1944 or who lived, but saw their comrades dying all about them, must have despaired of the purposes of their God. *However, the apparently deadly chaos of Normandy tells us something that even the most conservative churchmen are now admitting. That is, faith in a personal God can be deceptive, whereas faith in the purposes of a Creator serving the total needs of our miraculous life system can be more rewarding. A single person or a single society may not be fulfilled, and many may suffer in order that a larger purpose be fulfilled.*

If we will shift the scene from the individual anguish of Normandy to Allied Headquarters in England, the advancement toward a greater purpose can be seen. General Eisenhower, with a feeling of sorrow at the heavy losses, could note steady progress of the invasion forces and could sense eventual victory.

With the Allied advance and victory, the United States captured the most highly developed rocketry on earth, which was in the hands

---

[4] Pyle, Ernest I., *Here Is Your War* (New York: Henry Holt and Company, 1943; c. 1943, Henry Holt and Company, Inc.), p. 126.

of a German team. The director of this rocket team was Wernher von Braun, who was located at the vast rocket center at Peenemünde in North Germany, a prime target of the invasion forces. Von Braun had been a member of Germany's first rocket society in the 1930's and he had had many visions of the use of rockets for interplanetary flight. He was pressed into service of the military when Germany went to war, and his team had developed some fearfully effective rockets shortly before the Allied invasion. Most importantly, von Braun's team had finally developed the basic know-how that would make the launch of a machine from earth to outer space a reality.

Russia, invading Germany from the other direction, captured its share of German rocket and space technologists as well. The fruits of human desperation in World War II, the most advanced rocketry on earth, fell into the hands of the two most powerful nations on earth. The U.S. and the U.S.S.R., with a wealth of resources and skills, were the most logical places on all the earth to plant these vital seeds of knowledge.

World War II caused acceleration of the last and most important discovery by man, which would be needed for his travels to the stars. This was nuclear power. Refugee scientists, fleeing before the Nazis, arrived in the United States from Europe in the late 1930's and carried with them advanced information on the development of nuclear power. The survival of the United States would depend on finishing a nuclear bomb before Germany did. In a crash scientific effort, unknown in magnitude before that time, the United States found the means to release energy by a nuclear reaction. The power of a nuclear bomb was demonstrated in the nearly total destruction of two Japanese cities, Hiroshima and Nagasaki. The stimulus of war, as it had many times before, again brought forth a new energy source as the 1940's came to a close

World War II caused a cross fertilization of technology between friend and foe, and the relation between man and machine became even more intertwined over all the earth. A new scientific and material explosion became world-wide in the 1950's.

However, there came a time in the 1950's when it appeared that the great hope, humanity, and his symbiont, the fourth kingdom, would fail the life system entirely. The U.S. and the U.S.S.R. were engaged in making rockets to carry nuclear weapons capable of destroying vast areas of the world. It appeared that mankind was on the verge of ruining all its works in the next war. Man, whose pattern of war reached back to pre-history, seemed destined to direct all his energies to greater powers of destruction. Instead of turning our power outward to space, we were turning this power inward at the sacrifice and extinction of precious species of creatures on this earth. We began multiplying like the rabbits which were introduced into Australia without natural enemies. Whole areas of earth were becoming like high density poultry farms. Other forms of life were being snuffed out to meet human needs.

The reasons for the wholesale killing of the other forms of life were often agonizingly petty; to provide exotic foodstuff or to fulfill some obscure desire for decorations of feathers or fur. Man was not behaving as a creature of promise with a divine mission, but rather as any insect or reptile or bird or mammal which was suddenly given a power greater than that of his earthly competition. Humanity seemed on a path of destruction which could not be diverted.

The threats of nuclear bombs and missiles, and the runaway populations were accelerating as we proceeded into the late 1950's. Then, abruptly in 1957, launched by the momentum of this technological rush, the first artificial satellite, Sputnik 1, was fired into orbit around the earth by the U.S.S.R. The people of earth were awed by the sight of a new star roaming the dark heavens. It was as though a new bud had appeared on the tree of life, a new bud with the promise of spring, new life after the winter of gloom that had descended over the future of man.

In the years after 1957, the two greatest powers on earth, the supernations which men had been building toward for tens of thousands of years, launched space orbiting vehicles one after another. By the early 1960's both supernations had committed themselves to the beginnings of a space program.

A new living tree had grown nearly to maturity. It was the tree of the fourth kingdom, the symbolic growth of mechanical creatures, portrayed on page 61.

In 1962, Richard McKenna's memorable novel, *The Sand Pebbles*, was published. It is a novel about the adventure and eventual destruction of a fictional U.S. gunboat, the *USS San Pablo*. Cruising the Yangtze River of China, the *San Pablo* became caught up in the tidal wave of revolution that was the awakening of China in the 1920's. As if prophesied by this novel, a modern gunboat with a remarkably similar name, the *USS Pueblo*, too was caught up and destroyed patrolling the waters of the Orient.

McKenna recognized the phenomenon of man and machine, where the two spirits were inseparable. The central figure of this novel was Jake Holman, who joined the *USS San Pablo* as its engine room mate. He was thus described by McKenna: "Jake Holman loved machinery in the way some other men loved God, women and their country."[5]

Jake Holman's closest associate in the engine room of the San Pablo was a Chinese coolie, Po-han, who cared for and revered the engine as much as did Jake himself. Po-han was killed at the hands of a mob on the shore of the Yangtze River, and Jake Holman was grieving and reflecting on his death in the engine room of the gunboat:

> Po-han was still there, in the smooth, quiet stroking of the pumps he had rebushed, the quick, pulsing throb of the dynamo,

---

[5] McKenna, Richard, *The Sand Pebbles* (New York: Harper & Row, 1962; c. 1962, Richard McKenna), p. 11.

EVOLUTION OF THE FOURTH KINGDOM

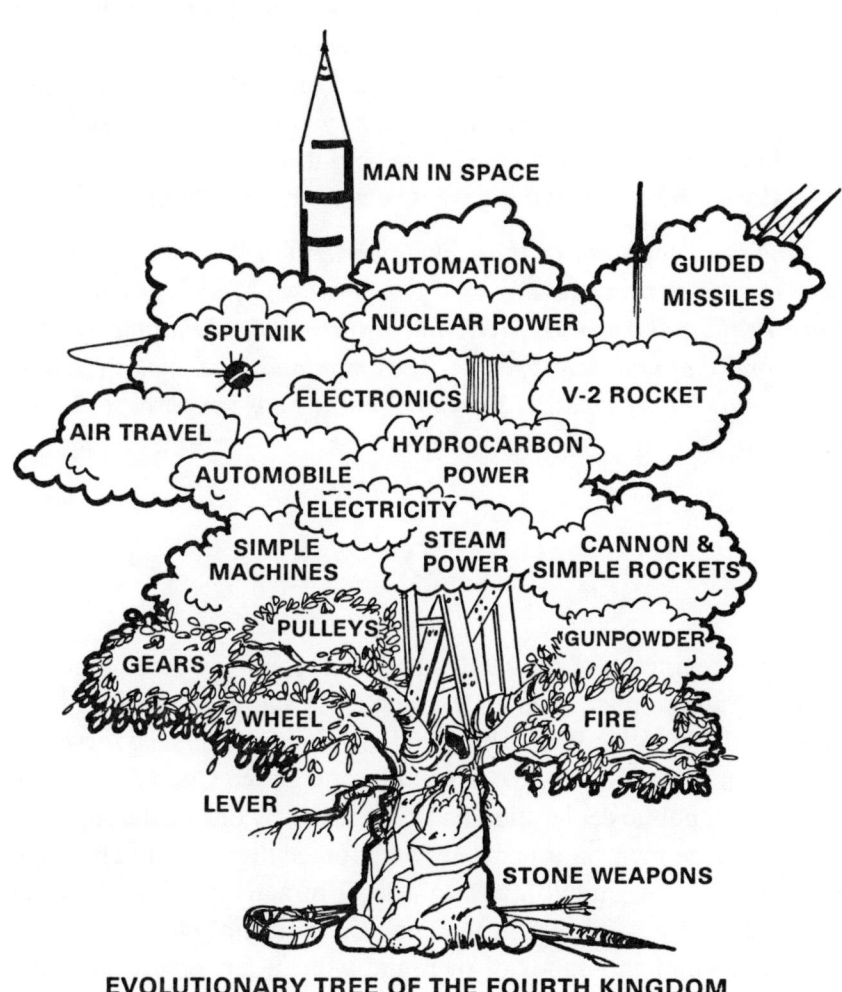

EVOLUTIONARY TREE OF THE FOURTH KINGDOM

the high, sculptured steel-and-brass gleaming engine. Deep in the foundations of the engine, Po-han was there.

He saw Po-han in the curling flames and heard him in the whispering steam and the trickle of water into the hot well. It all came from the sun and it went where everything went. Along the way it shaped itself so you could know it in a laboring engine or a warm and breathing man; you joined and mixed and knew. But you could not stop or hold it. It never ran backward. It went where everything went because it was everything.[6]

McKenna captured in those short paragraphs the essence of the flow of energy that was life, that was a moving machine, that was time. In those few words, McKenna also recognized that close spiritual bond that had developed between man and machine, a circumstance pursued by creation since man left his garden paradise in the forests of Africa.

At the beginning of the 1960's President Kennedy committed the United States to place men on the moon. He had been urged to meet this goal by his Vice President, Lyndon Johnson, who also served as Chairman of the National Aeronautics and Space Council. For the first time, a public declaration had been made by one of the supernations to place men on a heavenly body other than earth. The challenge stirred the United States to a level of activity which had not been seen before—except in the urgent necessity of war.

Hundreds of thousands of men and women left secure jobs and promising careers to join a fledgling organization whose purpose was to free life from the bonds of earth. No single corporation or governmental unit by itself was large enough or properly organized to handle the challenge. The United States National Aeronautics and Space

---
[6] Ibid., p. 365.

Administration (NASA) created a super organization of industrial corporations and governmental units to meet the immense challenge. They called it Project Apollo.

Those close to the operations of Project Apollo had never before witnessed such devotion to a cause and such feverish intensity of effort by so many people. A great many marriages came apart in cases where men had completely forsaken their women for their machines of space. The unbelievably smooth meshing of many organizational parts was a triumph of advanced management principles, said *Fortune* magazine at the time. *Rather could it have been that when man set out in the direction of his space destiny, he would find the smoothest path to follow?*

In order to place human history in proper perspective, consider the billions of years from the beginning of the earth until now as one long eternal year. Divide this eternal year into 365 "days" and 12 "months." Each "day" will be 12,500,000 years long, and each "month" 375,000,000 years long. Let us assume that the earth took form at 12:01 a.m. on January 1, the beginning of our first eternal year.

The earliest forms of life, primitive cells in the earth's oceans, did not appear until late spring or the summer of our eternal year. By the fall of the year, soft bodied marine animals appeared. During October and November, life began to grow in abundance on the sea and on the land; reptiles, fish and insects began to flourish.

However, even as late as the dawn of the last day of the eternal year, December 31, there was no sign of man. Not until 9:00 p.m. on December 31, did bipedal primates, the pre-men, make their appearance. Modern man, Homo sapiens, did not become visible until about twenty minutes before midnight.

With only twenty minutes left on the eternal clock, humanity began to spread rapidly over the whole earth. It was as though there was intense urgency. No new species had come so far, so fast. Civilizations did not appear until one minute before midnight. Now with each tick of the clock, we were changing the face of earth, and new mechanical creatures were emerging by the billions. The airplane, automobile, radio, electrical power and things we call modern were developed in the last *half second* of our eternal year.

The things that happened in the last few decades can be measured only in micro-seconds with our eternal clock. The burst of technology in the last fraction of a second was as a flash of light brought about by the forces of creation.

It was as the flash of light that launched the great rocket, Apollo 8, in December, 1968. In the last tick of the eternal clock, life from earth in the form of three men, left the gravity of this planet for the first time.

For the first time, living eyes from earth looked down on another planet and saw it as "near" and the earth as "far." The mechanical space creature and its human symbionts circled the moon. Most of the people on earth stopped and paused to wonder about this incredible event, a flawless flight to the moon and back. But by another micro-second on our eternal clock, most persons had forgotten and had again taken up their earthly pursuits.

In that one momentous event, the life system had taken the first beginning step toward casting its seeds to the unknown gravitational currents and winds of the Milky Way galaxy. Humanity was as mindless to the significance of the first Apollo mission as we think the dandelion and the coconut are, as they cast their seeds to strange currents of air and sea. However, all nature, and the power that sustains it, know of fertile ground where dandelions can grow, island beaches where coconut seeds can take root, and habitable planets where earth's life can flourish.

Frank Borman, the commander of Apollo 8, said that all the scientific knowledge that mankind had accumulated since its beginnings, from every country of the earth, had been needed to make man's first flight to the moon possible. *Might it not be true that all the scientific knowledge man had ever accumulated was meant to be used for man's destiny in space?*

Chapter 6

THE ARK

The ... idea of a nuclear-propelled space ark carrying civilization from a dying solar system toward another star for a new beginning was envisioned in 1918 by Robert Goddard. Possibly concerned about professional criticism, he placed his manuscript in a sealed envelope for posterity and it did not see print for over half a century.

Robert Salkeld,
"Space Colonization Now?"[1]

The persistent legend of Noah and his Ark is known to nearly everyone in Western civilization. A similar legend was well known among the peoples of the Middle East as long as six thousand years ago. The names, places and events varied as the story was passed from one generation to another, but the "story-line" remained constant—a worldwide flood occurred, and the only survivors were the animals and people who lived on a large vessel until the waters subsided.

At some point in time, the early Hebrew writers of the *Book of Genesis* wrote this legend down for posterity in the context of Noah.

A great deal of thought has gone into the debate on the truth of the

---

[1] Salkeld, Robert, "Space Colonization Now?," *Astronautics & Aeronautics*, September, 1975 (c. 1975 American Institute of Astronautics & Aeronautics), reference to a manuscript by Robert H. Goddard, "The Ultimate Migration," dated January 14, 1918, The Goddard Biblio Log, Friends of the Goddard Library, November 11, 1972.

story of Noah. The writers of the legend of Noah surely did not intend their readers to concern themselves with the literal truth of the story. There are huge, gaping holes in the story when it is looked at from the point of view of rudimentary science. And the people who wrote *Genesis* knew as well as an eighth grade biology student of today that the ocean teems with animal life that would not perish in a flood. Not only traditional marine life, but a great many mammals, birds and insects could live comfortably in a watery world. And of course a zoologist or transportation expert could not begin to imagine how Noah could have accomplished his feat with the resources he had available.

The writers of Noah rather were using a dramatic and intriguing story to cement in their readers' minds a truth. That is, all the living earth can be destroyed by immense powers beyond our control; in Noah's case the power was God. The story also shows that man and his works can save life from total destruction. The writers of *Genesis* accomplished their purpose, because the legend of Noah resides prominently in our collective consciousness.

While the legend is six thousand years old, only within the past decade or so has humanity progressed to the point where we could build an ark to rescue life from the very real hazards we face. Only now do we know enough about zoology and botany to attempt to keep animals and plants alive on an ark over a period of months or years. Thousands of men and women are now expert in the life sciences. For every expert in the field, there are many qualified amateurs, who take near-professional interest in the living habits of other creatures—a strange preoccupation for one species of animal, *Homo sapiens*. It seems we have listened well to the legend of Noah.

Unlike Noah, we will not be able to take every specimen of animal on board an ark that must travel to a new earth near a new star. We will need to select ingeniously just the right balance of animals, plants, bacteria, fungi and the like. We will need to design a precisely balanced ecology for the new planet, or the whole system of transplanted life will not survive.

At times when human arrogance cripples our perspective, we think our understanding of ecology will allow civilization to substantially improve on nature. However, nature maintains a balance in the midst of a complexity we barely perceive. *Everything we have learned or will learn about ecology will at best help us reduce the damage our civilization will do to earth's ecology. Our knowledge of ecology will find its ultimate fulfillment when it comes time to load the first space ark.*

Comic strips, science fiction books and fictional television programs all make a journey to another star look easy. The more popular writers have dealt lightly with the laws of physics in order to avoid the agony of working years-between-stars space journeys into their star-hopping adventures. However, a colony on a real inter-stellar ark will face the same kind of tedious journey our forebears endured on early sailing ships or transcontinental wagon trains, only for an even longer time.

Some fiction writers have certainly tantalized us with the glamour of space adventure, without telling us how almost impossibly complicated it all turns out to be. But just as well. No sense mixing reality with pleasure. But as we approach the day of serious ark-building, we will find a need to build a self-contained city in all respects, with the responsibility for this city to maintain its own gardens—food cannot be shipped in.

A reliable source of energy to power the ark will be imperative. There will be light and warmth from the sun only so long as the ark is in orbit around the earth during the construction and assembly phase. Once the completed ark sets off toward a distant star, the sun will quickly seem to diminish and become just another bright star in the heavens to the ark's passengers.

Two possible choices for energy are: nuclear fission using uranium as a fuel or nuclear fusion using hydrogen as a fuel. If a hydrogen fusion system were developed, the ark would in effect carry its own portable star or sun, since hydrogen fusion energizes most stars in the universe.

Once a source of energy is assured, a source of food is assured. In an agricultural section of the ark, a greenhouse with lamplight instead of sunlight can feed the ark's passengers. The passengers will survive on a vegetarian diet. Beans, peanuts, grains, potatoes, fruits and edible plants will be grown from high-yield, efficient strains of plants now known or being developed.

Clothing within the ark will be the absolute minimum required for function and not style. Perhaps a modernized loincloth, worn today by primitive tribes in comfortable climates, would be uniform for both sexes, with an optional halter top for women. Slippers or shoes may be optional. The most important items of clothing will be space suits to be used by the repair and inspection crews who must work outside the ark. Clothing suitable for the many possible climates on a new planet must be on board. Textile materials will be recycled, respun and rewoven over and over again. Synthetic fibers will be especially prominent. However, the ark society could also grow, spin and weave its own cotton, returning the used cotton fiber to the soil for natural recycling.

A whole spectrum of health supplies would be stored on board including: pharmaceuticals, blood collecting and dispensing supplies, vaccines and anti-toxins, surgical instruments and supplies, laboratory equipment and supplies, personal hygiene articles, and many other items to sustain health and preserve life for many years.

Some elements less critical to survival but essential to the comfort of the ark's people will be household and recreational equipment and facilities. Perhaps there will be room for a small exercise or game court. Everyone may otherwise need to be content with individual exercise equipment and competitive games of electronic baseball, hockey, tennis and football.

While most of the essential support items on the ark have been mentioned, and the associated design problems are mind boggling, we haven't progressed to the tough problems yet. The real research, design, and engineering challenges will come with the maintenance of an atmosphere, the creation of a gravitational force, the total recycling of every bit of matter on board the ark, and the caring for all forms of life for many years. The workings of nature taken for granted here on spaceship earth suddenly become monumental challenges for us when we try to duplicate a habitable environment in space.

The ark will begin its journey with tanks of atmospheric gases and excess water. When the ark is once underway, living processes will consume and create water, consume and create oxygen and carbon dioxide and create, at times, noxious gases, such as carbon monoxide and nitrogen oxides. An elaborate atmospheric control and processing center will employ the best we know of gas chemistry and air pollution abatement.

Life on the ark would be chaotic, if not ultimately deadly, without some form of gravity. The life support section of the ark may need to be in orbit around a center to create a minimum gravity. Perhaps the total ark will need to rotate like one huge disc or wheel.

If enormous amounts of propulsion energy could be developed, acceleration and deceleration forces could provide an acceptable substitute gravity. The velocity of the ark may then approach the speed of light and change the time dimension on the ark, i.e., a year on the ark may equal one thousand years on earth. The possibilities of achieving that much propulsion energy are so remote as to be well beyond the scope of what can be dealt with in today's reality, and thus the subject is outside the scope of this book.

Every drop of water, every speck of protein, every bit of material must stay in the living cycle. All wastes must find their way back into the system as pure water, clean air, and food. The ark's society will have no room for the cultural custom of burying its dead or even launching the dead to space, as the Vikings launched their dead to sea in boats. The bodies of the space ark's dead will become valuable protein, water, and minerals to be recycled into the living process.

The most humanity can ever know about zoology, botany, and medical science will be used to find a way to carry animals and plants on the ark. Plants, with their long lasting seeds, present the least problem. Grains buried for thousands of years with the Egyptian Pharaohs, have been shown to sprout when planted in the ground. However, animals will be carried in some state of hibernation or perhaps "frozen." More probably, animal eggs and sperm will be stored in some state of preservation. The first animals to come from the transported eggs and sperm will be grown in artificial wombs and fed by artificial placentas. When a colony is established on the new planet, the animals can then quickly revert to normal breeding habits for future generations.

By now, a great many of the scientific and engineering readers have blanked out from the mental overload involved in thinking about the problems to be solved. But, we'll go on.

All the equipment and all the systems on the space ark must be more reliable than the best airplanes and automobiles ever built. There are no service garages or maintenance centers between the stars. Service guarantees no longer apply. As we have already learned in our infant space program, reliability can be achieved by building two, three, or more systems into a spacecraft to serve as backup when the primary system fails. Thus the enormously complicated systems of the space ark will be in duplicate or triplicate. All the systems must be interchangeable.

There will be no need for the space ark to set off blindly for some unknown planet using untried systems. Telescopes placed on the airless moon can be made large enough to see planets orbiting neighboring stars. Astronomers will be able to select a number of likely candidates to be the new earth.

Our own solar system is a splendid practice and test area for space colonization. Mars is not ideal for earth's life, but pressurized stations can be established there to perform the trials of interplanetary life-transfer. The systems of a space ark can be tested en route to Mars. Means to transfer life from the ark to Mars can be checked out, a critical step in the process. It is one thing to find a possible new earth, and it is another to confirm it as "the place." The ponderous ark must first be put in orbit around a planet. Vehicles from the ark must be able to explore the planet without jeopardizing the rest of the ark. Finally, a "whole earth" colony must be transferred to the planet intact with as much support equipment as possible. An Earth-Mars practice of these steps will be a critical prelude to outfitting the costly, universe-ranging space arks.

It would be foolish to say how long an ark may be enroute to another habitable planet. There remain too many unknowns: the propulsion power, the mass of the ark, the distance to the nearest habitable planet-star combination. Neighboring stars are between 5 and 180 light years from earth or between 30 trillion and 1,000 trillion miles away from earth. Thus, tens or hundreds of years will likely be involved.

Children may be born and die without ever having seen mother earth or the new earth. Thus the space ark must carry a complete educational system. This system will be as important as the food, water and air to the ark's inhabitants. It was a carefully constructed civilization which created the fourth kingdom and which will ultimately create the space ark. The loss of this mechanically oriented, materialistic culture would mean the certain loss of the ark and its precious cargo of life.

A space ark, as a self-contained unit of life, will come as close as humanity can manage to duplicating the miraculous operation of a single living cell. When we have built and launched our first space ark, humanity then will have truly integrated its talents with nature. In the space ark, man will have approximated a seed, an egg or a spore and joined nature in its fierce desire to assure the continuation of life.

Not all the space arks launched from earth will find a promised land—a habitable planet. Not all a dandelion's seeds launched to the wind find a fertile spot of soil; not all the coconuts dropped in the ocean drift to a shore. The urgent needs of our living earth demand that many arks be committed in order to be certain of finding other suitable planets.

Billions of stars and their barren planets are the stepping-stones for perpetuating life across the whole of our galaxy. The colonies of people who reach habitable planets will recognize, by then, the human purpose of carrying the seeds of life throughout our galaxy. The space people will recognize, after they have achieved a stable population on another planet, that they may feel the urge to take another step across the voids of space, to yet another planet.

The colony may keep the earth-built space ark in orbit around the new planet for a hundred years or more if necessary. If the ark has performed well, the "new earth" people may simply refurbish and restock the ark in orbit and start off again in a generation or two, rather than build a totally new ark. However, human technologists being as they are, they will probably wish to make major modifications or build a totally new ark for the next trip.

The people of the "new earth" cannot expect to rest for an eternity on their planet. There are increasing findings by astronomers that most stars do not remain stable for nearly as long as our sun. Thus there will always be an undercurrent of urgency to keep moving, and spreading life to as many planets as possible in our galaxy. And with each new planet populated, the chance of life finding eternity in the heavens will be increased.

The chance of finding a new planet already inhabited is very small. Life evolved on earth because of a unique set of circumstances and conditions. The warm, gentle, watery, chemically rich conditions of earth are rare in this galaxy. We also are near a relatively long-lived and stable star. The sun gave earth billions of quiescent years and allowed life the time needed to evolve from simple fragments to elaborate animals. There is a growing suspicion that many life systems on some rare but promising planets were destroyed during their early stages of evolution. The radiant variation in their nearby stars created extremes of heat or cold that put an end to the vulnerable emerging life forms.

Life on earth has had time enough to evolve plants and animals that can withstand great extremes of climatic conditions. Creatures and vegetation from our space ark could easily flourish on a planet with conditions too harsh to support the creation and evolution of life through the more tender stages.

We can suppose yet another kind of planet where life had evolved only to the stage of single cell organisms. Further suppose its star or sun would remain stable for only another 200 million years. At best, only the very simplest organisms could evolve before life would be destroyed. But after a space ark deposits life from earth upon its surface, the planet would be filled with fully advanced, mobile life forms in a wink of cosmic time. Shortly, civilizations would arise and, 199 million years before doomsday, that short-lived planet would start sending its own space arks to the other billions of planets in the Milky Way galaxy.

Humanity now possesses all the skills and scientific information needed to build a workable space ark. We have the people available. Many world governments are wondering what to do with their excess people, whose labor becomes obsolete as our machines become more efficient. And we have far more than enough material resources. A hundred space arks would use an insignificant amount of material and fuel compared to what civilizations are using for their earthly pursuits.

A reliable space ark will consume all the excess brainpower with which mankind has been endowed, for the spreading of life is an ultimate purpose of the endowment. All nature is waiting for humanity to fulfill life's need. All the people of the earth are awaiting a purpose and a reason to unite in an effort which will bring dignity to their societies and nations. All our young are awaiting the chance to use their enormous brainpower instead of turning to abusive drugs to quench the fire in their unused minds or to relieve the anxiety of being turned away from their evolutionary destiny.

The legend of Noah says that the only humans to survive the flood were Noah, his wife, his sons and their wives. According to the story, then, we are all descendents of Noah. It appears the writers of the Noah legend suggested a symbolic truth. For the combination of all the people on earth and all their talents are needed to build a life saving space ark. Noah is a part of everyone's spirit, and the spirit of Noah is composed of all humanity.

Chapter 7

PLURALITY OF PURPOSE

> This might turn out to be a special phase in the morphogenesis of earth when it is necessary to have something like us, for a time anyway, to fetch and carry energy, look after new symbiotic arrangements, store up information for some future season, do a certain amount of ornamenting, maybe even carry seeds around the solar system.
>
> Lewis Thomas,
> *The Lives of a Cell*[1]

We have placed our greatest hopes in technology, yet it has given us our greatest fears. Weapons of war now exist too destructive to think about except for short, terrifying moments. The burning of fossil fuels to power our machines leaves residues of noxious gas in the air we breathe. Machines have given us a dominant position in nature resulting in a runaway growth of human population which threatens extinction to many other forms of life. We have such power that, if misdirected, it could extinguish the biological fire upon which our very survival depends.

Proposed solutions to these threats to society have differed radically one from another. Some suggest returning to nature and a simple, "native," no-technology existence. Other suggestions are that more application of pure science and technology, in a reasoned way,

---

[1] Thomas, Lewis, *The Lives of a Cell* (New York: The Viking Press, 1973; c. 1974 Lewis Thomas), p. 106.

will solve all the world's ills. New approaches to religion and philosophy come to the surface every day. The search for the meaning of life has become an obsession to increasing numbers of our young people.

The causes of our blindness to nature's plan for humanity lie within our own limited understanding of nature's ways. The groping profusion of nature is difficult to perceive or follow because of a principle we shall call *plurality of purpose*. Nature does not work wonders one at a time. Rather, the wonders flow down many paths simultaneously. The intertwining and interweaving of a great many acts of nature coordinate themselves and order themselves to meet life's needs.

We glimpse simple examples of creation's plurality of purpose in the world around us. We are seeing the true wonder of plurality of purpose as we probe more deeply into the delicate and complex ecology of our life system.

The human body provides mind-boggling examples of plurality of purpose, where a great many functions are smoothly blended into one small organ. The hand is particularly interesting. At the most primitive level, the hand is an efficient, mechanical device for grasping, holding or manipulating objects precisely, with a light touch or a crushing pressure. The hand becomes a weapon when it is shaped to a claw position where the fingernails can be used to scratch or dig. Shaped as a fist, the hand becomes a club. Fingernails turn from weapons to fingertip protectors when the hand is in an outstretched, vulnerable position.

The tips of the fingers are specialized sense organs in their own right. The fingertips are so rich in nerve endings for sensing pressures—the sense of touch—that textures and small objects can be readily identified by the fingertips without the aid of the eyes. The sense of touch is so capable of refinement that it has allowed the blind to read rapidly, through the Braille system, and to "see" objects. The hands also become an elaborate and critical part of the human love-making ritual, with an especially pleasurable versatility.

The hand is so sensitive to small temperature variations, that a wet finger held upright can be used to determine wind direction because of the varying degrees of cooling that occur around the finger in relation to its position in the wind.

The skin on the inside of the hands contains a pattern of parallel ridges to provide more friction for holding smooth objects. The skin ridges on the tips of the fingers are so varied in design from one human to another that they can serve as a positive means to identify one individual from millions of others.

The hands are a highly effective means for giving visual signals. The positions of the hands and fingers, their action and rate of movement quickly signal emotion and meaning. This ability has been used to provide deaf-mutes with a visual language.

The hand is claimed in some circles to have certain mystical purposes and powers which cannot be proven or disproven here. In any event, the hand described so far carries out a noteworthy set of functions. But we haven't yet come to the most important functions of the hand as far as the role of modern man is concerned.

The hand has become a critical link between the thoughts and images of our brains and human creations. The hand is the basis of the multitudes of our arts and crafts, and is vital to the homely necessities of sewing, knitting, embroidering, cooking, cleaning and on and on. It is therefore no wonder that studies of the brain have shown that control of the hands occupies unusually large areas of our brain.

It is the hand which must transfer the creativity of our brain to meaningful symbols on paper by moving a pencil, pressing typewriter keys, or by drawing an image. The hand must manipulate a computer to refine our thoughts in mathematical terms. The hand must hold the chisel or operate the lathe which converts machine parts from formless matter. The hand assembles the parts to make a machine. It is the hand that must act to control and maintain a machine, thus connecting the living thought processes to the power of a mechanical creature. Yet, the hand's higher functions must blend smoothly with all its basic animal functions, an example before our eyes of plurality of purpose in action.

Also we can glimpse the workings of plurality of purpose in the feathers of a water bird. Though they vary in size, all feathers consist of a similar structure which is remarkably simple for the jobs it must do. The large wing feathers scoop and propel air efficiently to give the bird flight. The wing and tail feathers can be bent and turned to give flight control and direction with delicate sensitivity. Body feathers provide exquisite streamlining to the bird's form to provide easier traverse through the air.

It would seem enough for the feather to carry out just those functions related to flight, but there are many, many more. Feathers are constructed to insulate a bird's body from heat or cold and to regulate the bird's loss of body heat in a very special way. By fluffing its feathers, a bird may entrap more air and create its own insulation, and by compressing its feathers, the bird will lose more body heat. When flying and creating excess body heat, the compression of the feathers gives streamlining and cooling all in one action.

Feathers protect a bird's body from the bites of insects or from sharp blows. The mat of feathers may save the bird from the claws or teeth of a predator. Feathers may receive a coating of oil which waterproofs them, giving the bird buoyancy so it may float or glide over the water with ease.

The feather goes on to provide intricately colored patterns for protective camouflage or for identification to friend or foe. The feather serves to color and enhance certain parts of the bird's body for sexual purposes. The single purpose of a feather as a coloring medium becomes a miracle of complexity by itself if we simply think about the annual color cycles of a mallard drake.

Thus, in a simple feather there is a plurality of purpose that is not instinctively comprehended. We normally observe instead only the very surface of nature's creativity at any one time.

The human mind has grasped so little of the plurality of purpose in nature that we have made some dreadful miscalculations in well-meaning attempts to "improve" on nature. Often what we thought was evil in nature, was good, and our good intentions were destructive.

For example, wolves have been feared and hated because of their predatory killing of game animals. Yet, carefully run biological experiments on Isle Royale in Lake Superior, have proven wolves help the population of moose upon which they feed. By killing off the weaker animals, the wolves eliminate defects in the moose herd's genetic pool, and the flaws cannot be passed on to succeeding generations. Without the wolves, defective genes could build up to the point of causing a series of "genetic collapses"—a catastrophic dying off of much of the herd. When the results of the Isle Royale experiment were complete, another of our conclusions about the workings of nature had to be discarded.

Nature, through plurality of purpose, weaves into the lives of the higher animals the task of caring for the reproduction needs of plants which cannot move about. It has not been easy. The patterns of behavior of the higher animals—the mammals, the birds, and a few insects—are enormously complicated. However, it is obvious to a careful observer, that nature has manipulated the life styles of the mobile animals so that they unwittingly assist the immobile plants. Some animals serve the sexual needs of plants by carrying pollen from one plant to another. Then other animals become deeply involved in carrying the resulting seeds about the countryside.

The relationship between the random growing, immobile plant and the highly organized, mobile bee is the classic example. The bee relies entirely on the sexual ways of plants—the plant's flowering and germination—for its food. Plants blossom at alternate times of the season to accommodate the bee's need for a continuous food supply, at the same time assuring that they, the plants, will receive proper attention in their turn. Many flower blossoms assume not only the shape, but the coloring of the insect to be attracted.

Animals and birds unwittingly carry about the seeds of life on the outside of their bodies. The seeds of many plants have hooks and barbs which catch in the fur and feathers of the mobile animals, to be dropped or scratched free at some new location.

Birds are attracted to the brightly colored fruit of many plants, an important source of nourishment. But imbedded in the pulp of the fruit are small seeds which have so hard a coating that they can survive the tortuous travel through the digestive tract of a bird. The unwitting role of the birds as carriers of new life to otherwise lifeless places was described by James Michener in his book *Hawaii*. Michener tells of an island's rise from the sea as barren volcanic rock and soil, lifeless in the midst of life's potential:

> The years passed, the empty, endless, significant years. And then one day another bird arrived on the island, also seeking food. This time it found a few dead fish along the shore. As if in gratitude, it emptied its bowels on the waiting earth and evacuated a tiny seed which it had eaten on some remote island. The seed germinated and grew. Thus, after the passage of eons of time, growing life had established itself on the rocky island.[2]

And like the Hawaiian Islands which lay barren in the sun when they rose above the Pacific Ocean, there are barren planets in our universe which could support life by the light of their nearby star. What then of humanity that has the talents to carry the seeds of life to the sterile reaches of our galaxy? There must be creative energy at work continually in our life system to adapt man to that role. But man is a much more complex animal, with an infinitely more elaborate job to do, than the usual seed carrying animals of earth. Thus, nature's task to involve humanity is much more pluralistic in its workings, and the resources needed to do the job are much more extensive.

Humanity's role as a cosmic seed carrier is so important that it has been, and is being, firmly woven into the fabric of our life, as the life of a flowering plant is woven into the life of a bee. Nature's plurality of purpose is continuously working within our culture today. As we

---
[2] Michener, James A., *Hawaii* (New York: Random House, Inc.; c. 1959, James A. Michener), p. 7.

look at human society in this chapter, we can see the continuous shaping of civilization to meet nature's desperate need.

AGRICULTURE

Agriculture, a systematic means to produce food, clearly separates man from the other animals of earth. With the development of agriculture some ten thousand years ago, civilization became a reality. As soon as there was civilization, men began to study the stars, and to wonder about the heavens around earth. Newly liberated people began to pursue new endeavors which we now call the sciences. The early scientists began to lay down principles of mechanics which were to lead to the technological revolution of today. With the carefully measured timing of plurality of purpose, machines and improved methods freed even greater numbers of people from the farm just as more were needed to fill new jobs in hardware technology and the sciences.

Only in an agriculturally sound nation could the phenomenon of our Apollo program have occurred, where 400,000 men were readily drawn from the population to work toward a destiny in space. Only in a future world with sound agriculture will millions be free to work on a space ark.

Agriculture served another of nature's purposes by reducing the growing human population's dependence on the lives of wild creatures for food. In order to survive, the early hunting tribes which populated the earth decimated the wild animals. As humanity expanded, all of the naturally wild animal life of earth would have been threatened with extinction without agriculture.

Hunting was a most important adaptation which carried man through a critical time in his development. However, as we become more trusting in agriculture as our sole source of food, the residual hunting instincts in us will fade. Then nature's spirit will again be fulfilled in a bounty of wild creatures with numbers as great as the ecology of earth can contain. No longer then will the rich animal life of earth cringe in the confines of a furtive existence, forever fearful of that powerful, meat-eating primate, man.

CULTURE

Urban civilization brought into being special living patterns we call culture: the enjoyment and appreciation of art, music, drama, dance and the special human interaction we call society. But what really is culture and what is its purpose?

Culture, stripped of its dignity, is simply a set of artificial substitutes needed to live outside of the satisfying stimuli of the natural world. Culture became a necessity as the human animal was removed from nature. Without earlier natural stimuli or the cultural substitutes, the human animal would become an emotional vegetable. People from the cities often refer to persons from the farmland or woodland as "uncultured," in a demeaning or disdainful sense. The point is, that those living closest to nature do not need culture for fulfillment. Rather in the cities, culture is not a luxury; it is critically needed for mental well-being.

It has been said by the noted longshoreman philosopher, Eric Hoffer, that it is in the cities where man has made his greatest advances. It is in the cities where a breed of people is becoming adapted to an entirely synthetic environment, devoid of any contact with the nature of the planet, earth. The city culture has replaced the stimuli of nature with the stimuli of art, theater, television, radio, music, sports. Many people living in the heart of our cities have come to ignore, in fact even abhor, a life in contact with nature.

Civilization today uses "cultural islands" which can move a human being great distances through a hostile environment in comfort. Cruise ships which spend weeks or months on the ocean are provided with a familiar cultural setting. The ship will have motion pictures, pleasing art objects, music, theater and other cultural substitutes.

Giant airliners traverse continents at altitudes where the air is so thin it is nearly space-like in its nothingness. To make the human passenger comfortable in this environment, there are art objects, music, in-flight movies, and even lounges to serve as social gathering places. All of these devices permit culture to support our mental well-being in such an unnatural place as an airliner.

Cultural developments are becoming more prominent as we move from nature at an ever increasing pace. Totally enclosed shopping and living areas have been built, and cities enclosed from nature under enormous domes are being planned. Great numbers of people will rarely need to leave a totally synthetic environment for weeks or months on end.

One cornerstone of higher culture is the practice of highly refined manners, an exceedingly delicate approach to relations with others, to avoid friction. This type of behavior makes living in close quarters of a city's society considerably more pleasant.

The culture and manners of city society will allow the human animal to live comfortably within the confines of a space ark, yet make it a bearable and survivable experience.

SEXUALITY AND SOCIETY

Human sexuality goes far beyond the needs of simple organic reproduction of the species. Sexual needs, in fact, so dominate the activities of the human animal that it is difficult to separate social structure from sexual requirements. Sexual manifestations are so prominent that it is not readily obvious whether civilization's structure was influenced by sexual behavior or whether a new sexual behavior adapted to the needs of civilization.

Zoologists, anthropologists, behavioral scientists and others have noted this fact during the enlightened scientific inquiry of this century. Among them has been Desmond Morris, a skilled and discerning zoologist who wrote one of the more popular books on the subject, *The Naked Ape*.

Both the male and female of our species contain features of their anatomy which cause sexual activity to be a very satisfying and pleasurable experience. The sexual organs of both man and woman are packed with nerve endings to give intense sensual feelings, far beyond that needed for reproduction. Further, the human female is able to achieve a sexual orgasm identical to that of the male in its nature and intensity. A full response by the female has not been noted in other animals, even among our closest primate relatives.

The human female is also unique with her prominent breast projections which are larger than necessary for milk production. Rather a woman's breasts function as sexual signalling devices, and the dense nerve endings in the nipples make them effective sites for sexual sensation and stimulation.

The fleshy protuberances of the nose and ear lobes are peculiar to humans, and the prominent lip structure of the mouth is not found in the other primates. These unusual features were hard to justify by anatomical function until studies of humans under sexual arousal showed that the ear lobes and lips and fleshy parts of the nose become engorged with blood and hyper-sensitive to the touch. These unusual features of human anatomy were found to be primarily erotic pleasure centers, features unknown to the other animals.

The naked skin of the human also provides a whole range of centers sensitive to the touch, giving additional erotic sensations. In most females, and in some males under sexual arousal, the naked skin will redden in distinct areas of the body which also serve as stimulating visual signals to the sexual mate.

Nature's efficiency deems that these additions to the anatomy are so specific that there must be an express and critical purpose. Sensuality in human sex relationships must have a vital role in nature's plan. Human sexual behavior is so different from that of other animals that its purpose becomes worthy of very serious discussion and consideration beyond mere mental titillation.

The most popularly held reason for the intense sexuality of our species is that of pair-bonding. This is the emotional cementing of a relationship between a male and female—romantically described as falling in love. Pair-bonding is thought to have been a necessary development during the million or so years that man had to exist as a hunting animal, from the time he left his primate-line existence in the trees to the time he began to develop agriculture. In a hunting society a strong pair-bond was built up through an especially rewarding sexual relationship between the male and female of a pair. This provided for a satisfying relationship and spirit of cooperation among all the male members of the tribe, since there was little friction due to sexual competition. A tribe required close cooperation of its male members if its hunting activities were to be successful. While the male was away on a hunting expedition, it was necessary for the female to be impressed well enough with this pair-bonding to remain faithful to her mate and to care for the children which resulted from this relationship.

However, pair-bonding is a factor in the lives of many other animals, including wolves and geese, which do not exhibit anything approaching human sexuality. Thus there must be more to the purpose of human sexual intensity.

In the early low budget days of zoology and anthropology, it was more convenient and certainly less expensive, to observe the behavior of "wild" animals within the confines of a local zoo. From these studies of zoo animals, some sweeping conclusions were drawn to the effect that most animals' motivations were sexually oriented. Aggression, fighting and socializing were all related to the acquisition of a mate, or mates, and to sexual dominance. The conclusions were correct for zoo-confined animals, but were wrong in general. Studies of the same animals in the wild, showed far less sexual orientation. Many of the animal activities which appeared to be totally sexual in the zoo became, in fact, matters of defending territory and food supply in the wild. In the real world of nature, sexuality played a greatly diminished role. While the earlier observations led behavioral scientists on a wild goose chase for a time, there was, nevertheless, an important revelation to come out of the error.

The findings said clearly that animals removed from the stimuli of a natural setting, and confined to an artificial environment, no matter how comfortable the surroundings, turn to increased sexual activity. When the excitement of the hunt, the defense of territory, the threat of predators, and the other stimulations of nature are lost, there is a critical loss of emotional stimulus. This loss is compensated for by increased emphasis on the one very basic animal response still available, the very emotionally stimulating activity of sex.

Starting from the time our earliest ancestors left the trees, until we began to build our first large cities of the Mideast, the humanoid animal was gradually removed from nature. *Therefore, there would need to be sexual compensation to balance our needs. And it seems creation did just that in the past millions of years. The structure of our bodies was altered to maximize the stimulus of, and response to, sex.* Archeology and history do not provide many good records with which we plot man's progress from the more casual sexual relationship of a primate to the intense activity of modern man. But there are a few traces. As soon as man began living in sheltered caves and in crude dwellings in pre-historic villages, there appeared all over the world carvings which were dominated by the figure of a nude, voluptuous female figure. These nudes were written off as representations of a fertility goddess.

Another landmark of man's sexuality was created with the rise of the cities. In these early cities, humans were entirely removed from nature and were in effect moved to a "zoo." The results were most predictable, for here was a creature that had evolved with a remarkable sexual apparatus, with all the accessory stimulating devices implanted in him. And once confined to a city, the usual compensating forces went to work. We would not have known of the sexual excesses of the city (relative to the tribal living of the day) if the moralists of the time had not condemned the sins and excesses and orgies of Sodom and Gomorrah, of Rome and Athens. We continue to find some very succulent pornographic literature that flourished as far back as human history can be traced. Sexually-oriented art, pornography to some, commonly adorned most of the ancient cities of the Mediterranean basin.

As materialism, technology, and the size of cities increased, so did sexual activity, far beyond that needed for the reproductive process. The conservatives blamed the sexual excesses of the early cities for their downfall; and to these moralists, we owe the descriptions of the debauchery, told in great detail. The increased sexual tempo was there to be sure, as certainly as the cities and their unnatural environment were there, but a careful analysis has shown no relationship between sexual activity and the downfall of the city as some finger-shaking historians implied. The sexual activity was more of a constant, and the cause of the downfall of the cities and civilizations was from other causes and effects, usually an invading army, a drought, the loss of water or a disease plague.

The technological age of today continues to give the city a sensual mystique. The "bright lights" of the city bring to youth instinctive visions of sexual promise and erotic pleasures. There is sex as well, in a more muted form, in rural America and in the backwoods. But sexual needs dominate a society to the degree that the society is urban, sophisticated and removed from nature.

Humans, then, have been physically programmed for millions of years to compensate for a life which would remove them from nature. While other animals make pathetic attempts in a zoo to achieve emotional balance through increased sexuality, their bodies are not devised for continuous sensual pleasure. Confined zoo animals, therefore, often lead neurotic and unhappy lives. Yet, a human male and female can be deliriously happy in an otherwise sterile city environment. Given the minimal requirements of food and shelter, the human couple can compensate for almost a total lack of everything else through the complex sexual pleasures they are capable of achieving. From this experience comes the common expression in our language, "living on love."

Life in New York City's Manhattan is for most of its inhabitants a life devoid of the usual sights, sounds and challenges of the natural world. The popular song by Rodgers and Hart, "Manhattan", expresses how love can make Manhattan a place of joy. The ends of the verses contain the key word:

> ... The great big city's a wond'rous toy
> Just made for a girl and boy,
> We'll turn Manhattan
> Into an isle of joy.

> ... The city's bustle cannot destroy
> The dreams of a girl and boy,
> We'll turn Manhattan
> Into an isle of joy.

... The city's glamor can never spoil
The dreams of a boy and goil,
We'll turn Manhattan
Into an isle of joy.[3]

The human animal can still find great happiness without sexual activity as long as he is immersed in the nature of earth. Thoreau, in his two-year stay at Walden Pond as a hermit, seemed to exude happiness, wonder and joy in nature when writing of his experience. Yet a male hermit, in Manhattan, too poor, old or ugly for continuing sexual experience, is often a desperately lonely person, an alcoholic derelict waiting to die, as a long parade of Skid Rows in cities across our country will attest.

There will be television and sound tapes on the nature of earth to keep the culture visually and intellectually alive on a space ark between the stars. But the human, as a recently evolved animal, would suffer considerable emotional unbalance without the touch, the warmth, and all of the classic sensations which accompany the varied experiences of the human love affair. This is not to imply a space ark must support a continuous sexual orgy for its inhabitants to survive, but there must be a heavy emphasis on love, sexual or otherwise. The relationships need be not unlike the emphasis given love and sex in the happier areas of our urban centers. But the emphasis must be much greater than among those who can daily enjoy earth's nature to fill the totality of their waking hours.

Thus, the plurality of purpose of creation has thus shown yet another face in the intensity of human sexuality. The space ark will remove man completely from the nature of earth, yet not much more so than "the boy and girl in Manhattan." It will be largely their specially constructed sensuous bodies and minds that will give humans a richly rewarding substitute for the sensual experiences of a nature left behind on earth.

---
[3] Rodgers, Richard and Hart, Lorenz Milton, "Manhattan," c. 1925, Edward B. Marks Music Corporation.

The human animal seems to be especially flexible in adapting his sexual needs to the environment of his society. One extreme example is in the mountains of Tibet where as many as five or six farmers share a single wife in order that the birth rate may be kept in balance with the available land. In the Middle East, the taking of multiple wives is commonly practiced and accepted in social and economic circumstances where the involved parties and society itself tend to benefit.

We will be faced with many complex social-sexual problems in the micro-culture that must survive intact, with a minimum of friction in a space ark over years, or generations. The answers to these future problems may lie with new cultural experiments which have sprung up in the United States, the new communal societies. Remarkably, the communal society experiments started voluntarily, from the grass roots, coincident with the first serious efforts by our technological society to send spaceships beyond the earth.

The emergence of communal type sexual-social living is not restricted to one pattern. Rather, as evolution would dictate, communal societies have taken almost every conceivable form *"in order to try everything to find everything."* [4] Some communes consist of nothing more than clusters of married couples sharing common living quarters, mostly for economy's sake, a practice at least as old as the Great Depression of the 1930's. However, the classic commune is a very loosely structured group of individuals who live together in a large, rambling old house. No great care is taken to assure a balance of male and female members, but rather a care that there be a balance of individuals who can live together in harmony. Not only are the duties of the household shared, but the children are somehow well cared for in the adult dominated commune. Some of the male and female members are legally married and may remain faithful to each

---

[4] Teilhard, op. cit., p. 110.

other. Ranging to the other extreme, there may be unmarried individuals with polygamous or monogamous sexual behavior with wide variations. Other individuals in the commune may withdraw from sexual activity, at least temporarily, for one reason or another.

But what is important in a successful commune is group harmony, shared duties, special care of children and an atmosphere totally satisfying to varying sexual needs. The commune can also fulfill the human need for love and for belonging. A proper commune can work without the disturbances of jealousy, hate, and other antisocial evils which all too often accompany intimate group living.

And so, in the United States, as it enters a technological age capable of space exploration and travel, there are evolving thousands of groups experimenting with hundreds of combinations of communal living. Groups are dissolving and reforming along new lines. The communal groups are searching for the best micro-society within which members of both sexes can live in an intimate way.

Communal living is worthy of our tolerance. More so, it is worthy of serious study. For long before the first individuals are selected for a journey to some distant planet, there will have been considerable experimentation with communal sexual-social groups. Some ideal or near ideal situations are now developing if we will only look for them. A social group will evolve a life style from the confines of a city dwelling which will fit the confines of a space ark.

The stimulation and communion of human love and sexuality and the warmth of a well-balanced society will keep the human spirit alive in the sterile environment of the space ark, in the bleak voids of space away from the sun or earth, or any star. The human spirit has survived over the generations to persevere in city environments far more oppressive than a proper space ark will be. It has always been, and always will be, some blend of love and social custom that keeps the human spirit alive.

MINIATURIZATION

The faster, more reliable automobiles of today required the practical application—the "living" existence as it were—of many models or "generations" of lesser quality vehicles. In each model, there was the production of millions of automobiles. Through this "living" experience of multitudes, ideas and designs were tested and found worthy to carry forward to the next generation of automobiles. Similar parallels can be found for all of our mechanical devices: radio, television, airplanes, ships and the rest. The similarities to animal evolution cannot be missed.

Only a supernation, technologically, could support the proper evolution of mechanical life. There had to be thousands of industries producing millions of mechanical devices which would provide for a "living" world of machines. A sound economy allowed millions of consumers to purchase and use automobiles, airplanes, radios, television sets, motorcycles, trucks, measuring instruments, calculators, pumps, medicines, washers, dryers, furnaces, tools, preserved foods, houses, and all of the thousands of other items which make up our modern standard of living. The testing and improvement of these everyday items directly contributed to the mechanical reliability demanded for space flight. The new ideas and inventions that have come about through the space programs have been fed back into the mainstream of science and industry. The use of space technology on earth, in millions of consumer products, directly helps the space effort. Millions of users on earth test and prove the reliability of systems that cannot be allowed to fail, out between the stars. Safety activists may not realize the importance of their concerns about product reliability to our space efforts.

The latest "models" or species of animals compared to, say—dinosaurs—tend to be more efficient in their use of power, in the space their bodily functions occupy, and in the multiple use for every part of their anatomy. As birds evolved from reptiles, all the life functions were not only improved upon, but were forced into a more compact and lighter form. This gave birds the mobility to assure the presence of life in every corner of this earth.

Great power must be squandered to move each pound of matter away from the gravity of earth and to move it toward a new star at great speed. Therefore, machines and their accessories have to be miniaturized if they are to be useful in space. At the beginning of the Industrial Revolution, the first steam engines were bulky and ponderous in relation to the jobs they could perform. Since then, we have moved quickly toward lighter, more compact, and yet more powerful machines. The difference between the bulky steam engine and a compact jet or rocket engine of the same horsepower is astonishing. Radio, television, computers, and other electronic devices critical to a space effort have been ingeniously miniaturized in the last decade, coincident with the first ventures into space.

As we reached a means to cast the seeds of life into space, miniaturization effort reached a crescendo with enormous bursts of creativity and in spectacular technological breakthroughs. Certainly, miniaturization has other benefits to society and culture. That is to be expected if creation's plurality of purpose is operating normally. But the net result is an ability to package the hardware and the seeds of life in the lightest, smallest package that can traverse the universe.

SCIENCE

The greatest achievements of science pale when compared to the accomplishment of a life system that existed hundreds of millions of years before man. The best of our scientific efforts would not allow us to invent the simulation of even a single living mosquito with all its functionings. But we continue to delude ourselves. In this past decade, men working on the outer fringes of the life sciences were able chemically to take apart a living virus into two lifeless components. The two "dead" chemicals were mixed together and they became "alive" again. The newspaper headlines boasted that man had "synthesized life." However, this achievement, as important as it was, had no relationship to truly synthesizing life. In contrast to the real workings of the life process, this act was like that of a monkey that discovered he could screw a nut onto a bolt and then claimed he had invented the nut and bolt, when in actuality he had no part in mining, refining, alloying, forging or machining the steel.

And, so it goes for most of the rest of science. Science consists mostly of discovering and cataloging the natural forces and wonders of creation that have been in operation for billions of years. Science is simply the putting together of pieces of a puzzle that creation has already constructed.

There are some areas of science in which man has done unique things. We have discovered ways to put immense energy into machines. A huge vehicle can be lifted into the air and made to travel faster than the speed of sound. We, therefore, feel we have exceeded the performance of the birds. But our machines are not nearly as efficient in their use of energy as the birds. They are not nearly as adaptable, nor as maneuverable. The secret of an airplane's performance is its ability to consume great quantities of fuel to develop great power. But the airplane's fuel is organic, provided for us through the living and dying of nature's creatures that stored sun energy in their body tissues. The airplane has been a useful proving ground for the high performance machines we will need to leave this planet. But the airplane does not even approach the compounded miracles which comprise the daily living existence of a single bird.

Without medical science, rampant disease plagues would make close-living, large populations hazardous, if not impossible. Longer life, made possible by medical science, makes greater technical achievements possible in a single generation. And the best of medical science will prevent a total catastrophe from overtaking the population of a space vehicle in transit for years between the stars.

Improvements in medicine have created an unnatural situation on this planet. People with serious genetic defects and physical malformations live normal lives with special medical assistance. This has done much for humanity, to keep as many people alive as possible. However, genetically defective people are free to live and to produce offspring, a practice which, in turn, spreads this genetic defect even further through the population. Given enough time and adverse circumstances, there will be a "genetic collapse" and the dying off of great numbers of the population.

On the surface man's medicine seems to be fighting a natural selection process which will assure a healthy population. But if we again look at man, not as a creature to be optimized for this planet, but to carry life beyond earth, this medical anomaly (or future disaster) has a useful purpose. Medical science has had a real life laboratory of defective populations to study. A spaceship which may travel years or generations to reach a habitable planet, will have a small, fragile, carefully controlled society of humans. A genetic catastrophe, nature's answer to breeding carelessness, cannot be allowed to happen in this small society traversing space. Genetic problems can be avoided by what has been learned about the causes and detection of genetic defects in this real-life laboratory on earth.

Medical science has also provided another means for man's biological separation from nature. Methods have been developed to easily control population. Other creatures breed to the limit of their food supply or until social tensions become unbearable with overcrowding. But we, again, have separated ourselves from nature in the control of our numbers.

The greatest apparent anomaly, again, is that birth control techniques were developed and practiced first by the most advanced technological societies, those which can best afford to support higher populations. The apparent confusion in purpose fades away if we again note it is from the technological societies that our space travelers will emerge. In the confines of the limited world of a spaceship, the casual earthbound practices of birth control suddenly become *critically necessary* so that the balance of life can be maintained.

Creation's plurality of purpose works also for that society which employs birth control for earthly purposes. A nation will not maintain a strong technological position if its number of people gets out of control. All the nation's energy would be turned toward supplying food for survival of its massive population. The quality of life, which we equate with material and technical benefits, would diminish. And the ability to support a space effort would disappear.

Power was given the human animal to lift himself above all nature, to free himself from the bounds of earth, to reach the stars. Technology has also freed the human animal from nature's population control—disease and natural enemies. Therefore, if there is nothing to control his numbers, space capability will fade, and the resultant stress of worldwide overpopulation will cause an inhumane chaos beyond the most horrible consequences ever imagined.

One other earth science unique to man is astronomy. This includes the study of the physical laws of the universe and the precise measurement of the positions and movements of planets and stars. A newer and related science, astrophysics, is beginning to understand the birth, life and death processes of stars. How astronomy applies to our mission in space is too obvious to discuss further.

COMMUNICATIONS—IDEAS ACROSS TIME AND DISTANCE

Communication abilities among the animals appear to be only as complex as necessary to adapt to the environment of earth. Nature would not deem otherwise. Any one creature does not carry more communications equipment than needed, a law of evolutionary efficiency. Animal communications are limited to important earthly reasons: identification, sexual attraction, expression of territorial rights, signalling the presence of food, and signalling danger.

Humans employ techniques of sight, sound, and odor to communicate in the same way and for the same reasons as the rest of the animal kingdom. We often give our visual signals as instinctive subconscious gestures. Literature appearing in recent years teaches how to read these primitive signals in others, to obtain better insight into human relationships. A grimacing face with parted lips, bared teeth and an animal snarl, is an obvious gesture of aggression. Less obvious to many is the female's use of raised eyebrows followed by downcast eyes as a signal of sexual interest. Dogs, who have joined man's society, have retained their animal ability to read visual signals. Dogs are so proficient at this that their masters often come to the conclusion that the dog is "reading their mind," not recognizing that the human is "talking" to his dog in great detail by subconsciously projecting visual signals continuously.

Humans use grunts, sighs and screams in the same way animals do. And the enormous sale of underarm deodorants and soap has been our civilized way to prevent our built-in odor communication devices from working.

While retaining our primitive animal techniques, man has progressed far beyond all the rest of the life system in the time, energy and variety he puts into more complex communications. Since nature's laws do not allow for purposeless abilities to evolve in a species, it will be fruitful to analyze man's vast communications ability. Man has expanded his range of sounds, by the use of words, so that he can not only express animal emotion, but can *describe* emotions. Man not only shouts a danger signal, but he can *describe* a danger. Man not only makes an animal sound of finding and appreciating food, but he can *describe* the food or any other important object he has found. The significant portion of our brain devoted to speech can only be justified for a creature that would find it necessary to communicate with others of his species in some other way than face to face, where one person is remote from another or where a few individuals are remote from the society in general. The human vocabulary provides the ability to describe an image or event to another human out of the range of sight or sound or smell.

Closely allied to enlarged human vocabulary is our unique ability to transmit ideas by writing down meaningful symbols to be read. Here we have the skill of reading and writing. Again, the major purpose of this skill is the transmission of ideas, emotions, and descriptions over some distance, where one being is remote from another. Writing is also often used as a means of communicating with and coordinating masses of people. But the written symbol has the added value of communicating over long spans of time. History would not be possible without written symbols. Technology could not have moved forward across many generations without the written records of ideas.

One may well argue that our communications have many more

purposes than giving humanity the faculties for space travel. One could insist that our whole technological base, which is civilization, would crumble without our communications system. But then, it seems that civilization and all of man's technology is pointing toward space. Likewise, one might argue that any degree of order on earth would crumble without communication skills. That too, as was mentioned earlier, bears on the premise that coordinated efforts between great numbers of people are required to make possible any huge technical effort. The small, fragmented Indian tribal societies of the early United States and other parts of the world enjoyed a quality of life different, but certainly not "worse" than the highly communications-conscious society we live in today. The low communications level of a tribal society has been and is far more in balance with the nature of our planet earth. But, there is no way mankind could have flown to the moon and back without using every communication tool we possess.

As the distances become greater, communications become more difficult and costly. Many of us have met the hard realities of this problem by trying to choose the right words to keep a telegram as short as possible to save the money of transmission. Yet, we choose the best words to insure that the meaning of our idea or emotion or fact will be understood by the person receiving the telegram. As we approach the challenge of communications across space, we face the problem of transmitting words across millions of miles, where there may be hours, days or years of delay in a round trip message. Every bit of data thus transmitted must be fully meaningful.

It has been interesting to watch our language compensate for the cost of transmitting words over great distances. In the 1930's and 1940's we began to shorten the identity of governmental and business organizations by our "alphabet soup" approach. The Tennessee

Valley Authority was better known as the TVA; the Works Projects Administration, the WPA; International Business Machines became known as IBM. In the utterance of a few letters, a complex organizational concept could be transmitted rapidly between humans.

As the world entered the time of our first space exploration, the "acronym" was inserted into our language. The acronym is simply the coining of a new word to explain a complex technical or organizational concept. One of the earliest and most widely known acronyms is "radar," derived from its technical description "radio detecting and ranging." The memory magic of the acronym is that it is usually structured from the first letters of a series of words describing the item or organization. The word "radar" itself, reading the same whether spelled backward or forward, also mimics the action of the radar signal which is broadcast as a radio wave and which bounces back from an object as the echo of a radio wave.

Appropriately, the identity of the National Aeronautics and Space Administration has been shortened to the acronym NASA, (pronounced naa-sah) where the acronyms found their greatest and most practical value. Words in the Apollo program demonstrate this. The Apollo Lunar Surface Experiment Package, whose workings could only be described in a short book, is simply known as ALSEP, a short two-syllable word which was uttered many times between moon and earth by the Apollo astronauts and Apollo ground operations. Acronyms, transmitted as one or two-syllable words, easily understood and recognized across the abyss of space, have done much to aid in the success of our space exploration where communications are critical.

Critics have accused urbanites of an excessive use of words, of too much conversation, of useless oceans of drivel. But again, evolutionary forces are working to expand a skill far beyond its apparent earthly usefulness. The "hot air" of words bubbled through our society has given us a distillate of language skill of the highest order. Out of human words has come a space communications device of the highest efficiency, a few sounds encompassing the understanding of complex ideas and facts.

The final keys to unlock the human communications potential were the inventions of radio and television. Radio and television are well accepted as entertainment or amusement. It is peculiar that any animal could be so fascinated by hearing a voice or seeing an image transmitted over a great distance that it would prefer this means of communications to talking or seeing others of its kind first-hand. Yet, man has an almost fanatic interest in remote, electronic communications and gives highly paid homage to those who so communicate with us—the "stars" of radio, the motion picture, and television.

It seems as though Homo sapiens, we, as a species, are programmed instinctively to adapt to remote communications. How many times will our children prefer to visit with a friend by phone rather than in person? No others of nature's creatures would prefer electronic images and sounds over live representatives of his species and the stimulation of nature. Can you imagine a bird leaving its place in nature to spend any of its waking hours watching other birds on television?

Again, through evolutionary purpose, humans are adapting to a life where remote communications can be entirely satisfying. People are adapting for the time when a space ark will be in transit between the stars for months, years, or lifetimes. All the ark's inhabitants will maintain their culture through the medium of remote communications.

Hopefully, there will be "live" contacts with earth as the space colony traverses the stars. But as the distance of an ark from earth extends to one or more light years, the time for transmission and the weakness of the signal will create difficulties. Here our magnetic tape technology provides the solution. The storage of sounds and sights and data on miniaturized magnetic tapes will provide our space society with a whole library of learning and culture. Educational courses and entertainment can be stored on tapes to be televised to adults and to the children born on board the space ark or on the new planets and thus retain our culture intact.

Humanity's technological burden, the whole kingdom of machine creatures man has brought into the world, can be carried on the space ark as a total encyclopedia of all pertinent knowledge. The equivalents of libraries full of blueprints and instructions can be stored in a relatively small space with magnetic tapes.

Tapes will store total information on how to build a new ecological system of life on the new planet from the carefully stored seeds and fertilized eggs brought across the vastness of space. Children would be trained to Ph.D. levels of botany and zoology from televised courses taught them from the same tape communications devices. Children of our space ark pioneers would benefit from television tapes prepared by the foremost scientist-teachers of the world. Their lot would be no different from those learning from television today, from the pre-elementary programs such as Sesame Street, to the thousands of advanced university lectures.

Humanity's network of communications skills and devices can be summarized as nature's means to carry technology across many years or countless miles. Nowhere in nature is this skill needed. For nature carries forward the whole of life on earth, through genetic codes and similar means, far better than man's greatest inventions or historical activities have ever done. Man, however, is needed to transmit a technical culture and an ecology across light years of space.

RACE AND ADAPTABILITY

Homo sapiens has made many physical adaptations to fit the extremes of climate on this earth. We call these racial differences. Yet, in spite of specialized physical changes, we have remained as one single species of animal, capable of interbreeding—a phenomenon which disputes the laws of evolution. Normally, an animal establishing its habitat among the far corners of earth would have branched out into many new and specialized creatures. Zoological history would predict that man should have branched out into many species by now. But man, with all his racial differences, remains as one single interbreeding animal.

Civilization and technology would not have been possible if man had been fractionated into a number of different species. But the most important benefit of racial differences will become apparent when man seriously plans for his space arks. It is probable that after searching the heavens for habitable planets, we will need to make environmental compromises in selecting a planet to colonize. It is unlikely that the climatic variations of a distant planet will be arranged like those on earth, with tropical temperatures at the equator and cold, but livable climates in the upper latitudes. Rather, a new planet is apt to be much warmer or colder than earth, with vastly different levels of light intensity from its star.

The equatorial regions of a cold planet could possibly have a climate, at best, like northern Canada, Siberia or the Scandinavian countries. The light will be weaker than the earth's average. In such an environment, relatively light-skinned people will adapt better than dark-skinned ones since ultraviolet light, required for sound bone structure, is best absorbed through light skin when the light source is relatively feeble. Lighter skinned people are most apt to survive on such a planet. If the planet were not only cold, but windy, then the addition of a fold over the eye, characteristic of Asian peoples, would be desirable. The fold, giving the eye a "slanted" appearance, protects the eye from freezing. This racial characteristic was important to survival with cold winds sweeping the Asian continent. The Lapps who inhabit the northern regions of Scandinavia have made the same adaptation.

On the other hand, there will be warmer planets whose climates, even at the polar regions, will be as hot as earth's equatorial regions. This also implies stronger lighting and higher ultraviolet radiation. On this type planet, the darker skinned humans will have an advantage. The Negroid racial characteristics give a person the ability to survive a hot planetary environment, with high light intensity. The dark skin pigmentation will protect the body from too much radiation, avoid skin cancer and yet absorb enough ultraviolet light for good health. The tightly curled, heavy hair on the head protects and insulates the brain from overheating and radiation. And a completely hairless, dark body is most effective for throwing off excess body heat.

Chances are that there will be uncertainties about the climate of a planet which a space ark sets out to colonize. It is probable that space arks will be sent on their way furnished with telescopes and instruments which will be used to select a planet only after the ark reaches a star thought to have a planetary system around it. Thus, a broad mixture of human racial types would be chosen for the journey. In this way, the space ark society will retain a broad mixture of human racial characteristics in the gene pool.

After a new planet is colonized, the best racial characteristics necessary for survival will become apparent. Natural environmental factors will determine which racial characteristics will survive in future generations. The second and third generation "new planet" children with the racial characteristics best suited for the new climate will have distinct advantages. Thus a broad mixture of racial characteristics in the colony's gene pool will give the human contingent of the space ark its best chance for survival.

On the planet earth, the inability and unwillingness to accept people of a racial type different from our own, has been the cause of prejudice and strife. Yet, for interplanetary purposes, racial differences and humanity's ability to interbreed become important biological adaptations. It may be that the barriers that exist between the races, will continue to create friction in our society until man recognizes his space destiny. Then, we may *honor* and *respect* the wonder of racial differences in humanity. Perhaps the racial tensions we are agonizing over are simply a demonstration of a purposeful instinct in man. There have been many well-meaning attempts to totally integrate the races in the United States. These attempts, for the most part, have met with opposition from all races. It turns out that a minority race really wants recognition and does not want to be lost in the "melting pot". What the races want is equal social consideration, equal justice, equal job opportunities and equal respect. Surprising to many sociologists, minorities have fought to maintain racial distinction, once equality was reached. Our better social instincts today mirror the society that must be maintained on a space ark. Racial differences are of great value if the most important mission of our being is to be served.

It is another tribute to humanity that we have created useful "racial" differences in our machines. We have created machines which function on dry land, on water, in the desert, on snow or ice, or in the humid tropics. Man has specialized his machines so that he can choose the type best suited for his new planet's climate. The success of the snowmobile was originally due to the public interest in sport and amusement. But life will never again be so hard in the northern latitudes of earth, with the mobility of snowmobiles available to so many people. And future generations, finding themselves on a cold planet, will be thankful that their forebears left microfilmed instructions on how to construct an electrically driven snowmobile.

NUCLEAR FUSION

In the past few hundred years, we have learned to apply energy to give machines great power, to heat and cool our environment, to communicate across great distances and to give light in the darkness of night. The source of the energy has been oil and coal, the residues of the bodies of billions of living things, which stored up sun energy in their life processes.

It took millions of years of their living and dying to create oil and coal. We have been using up these irreplaceable fossil fuels at an ever-increasing rate and we can already see the time when our traditional energy sources will be gone. We are simply going to run out of coal and oil.

Fossil fuels are only temporary energy sources. The fossil fuels seem to be a storage battery of sun energy whose charge can power man's technology for a relatively brief span of time. Then we will have to use the sun's energy directly for our power. Or we can create our own nuclear power source here on earth.

Humanity, and just in the nick of time, is close to creating a source of "sun power" on earth. We now have the ability to control the hydrogen fusion process, which powers the sun as well as the destructive doomsday H bomb.

This energy from controlled nuclear fusion is needed long before we consume all our oil and coal since they are invaluable raw materials for much of our industrial technology today. Burning all our oil and coal for pure energy will be catastrophically wasteful. Burning fossil fuels for all our increasing energy requirements also raises the grim spectre of runaway air pollution, with many unknown hazards to life. Uranium fueled nuclear fusion reactors are good only so long as our uranium holds out; and we can easily project the end of this power source as well.

Controlled nuclear fusion is expected to become a reality in the next twenty-five years. It is not surprising that man would discover nuclear fusion as a source of energy. The images of life all come from the radiant power of the fusion reactions of the sun. The images of humanity are only possible by the energy of nuclear fusion pulsing through the body of the life system. Therefore, it would seem a certainty that man would someday discover a way to produce the energy of his being.

All the rest of nature is content to live within the limits of the energy of nature's food supply, the daily allotment of sunlight. Sunlight is sufficient to power the plant kingdom's food-making powers and to supply the food for earth's animals. Man, if he were destined or content to live as one of earth's animals, would not require additional energy sources. But man is destined to carry life beyond the earth, and immense power is needed to leave the gravitational pull of earth and the solar system. And more importantly, a space ark will need its own source of power, its own sun, to power life on the journeys between the stars, where natural radiant energy is too feeble to support life.

Nuclear fusion reactor designs as visualized today are awkwardly large. But our first designs of energy sources always seem to be excessively large. As our usual inventiveness works on the nuclear fusion processes, it is a certainty the reactors will be miniaturized enough that a design will be practical for interstellar travel. A space ark powered by a nuclear fusion reactor and a modest quantity of fuel could operate in space or on a new planet for thousands of years.

Nuclear energy will help assure life's survival on a lifeless planet which will have no fossil fuel to start with. The new earth will have only the radiant energy from a nearby star, its "sun." But if man is to exist with technological comforts, he will have only nuclear energy to power his machines, to warm him, or to light the darkness until trees and other plants become established. So, among the most precious microfilmed instructions that the first humans will take with them to a promising new planet, will be a means to build a nuclear fission or fusion reactor. Hopefully, the elements of the new earth can provide the necessary raw materials.

Modern civilizations will decline without fusion power. If man chooses war as his outlet for creativity and aggression, nuclear power will destroy him. If this same creativity is turned to space, life will survive a long journey with the benefit of energy. A nuclear reactor will be humanity's salvation on a new planet.

EMPLOYMENT AND SOCIAL WELFARE

Industrial automation has furnished a high standard of living for most people and a rising productivity for every working person in the United States. But there is the other side of that coin, and the other side is a distasteful social problem settling on our collective consciences. There are growing numbers of people permanently unemployed; there are second and third generation children of the unemployed who are also idle, receiving their sustenance from welfare as their fathers and grandfathers have done. The welfare problem has been rapidly increasing in the last decade.

The premise of the fourth kingdom shows that man's society is structured to free more of its numbers to pursue science and technology and its ultimate mission to the stars. If that be the case, then we should see in a national space program a solution to unemployment and its attendant welfare miseries. And so it seems to be. For in the early 1960's as the United States worked feverishly to place men on the moon, our nation simultaneously enjoyed a dizzying climb in prosperity. Then in the late 1960's, the United States diverted its spending and efforts away from space exploration. Hundreds of thousands of loyal space workers were suddenly unemployed. We created a permanent depression for a great many of the unemployed workers, and for thousands of small service businesses which indirectly supported the space program.

The unbelievable spectacle of space workers, who had accomplished modern miracles in their mission, standing in lines waiting for their welfare checks, filled the pages of the mass media. These men, who had received government funds to do the most creative work man had ever done, were now receiving governmental funds for doing nothing. Imagine the grim irony which lay upon those million men, who had basked in the glow of the success of their work when the first humans stepped upon the moon.

An enormous amount of work by millions of people must be done here on earth before the first space ark can set out. Total communities of people could be employed on just one of the myriad of design and operating problems involved in the space ark. The pride of accomplishment of the Apollo program could be multiplied many times. And humanity will have found a high purpose which will extend to the furthest reaches—or depths—of our welfare problem.

PLURALITY'S ARROW

As an Indian shapes an arrow of formless rock, we too have chipped away at the apparently formless shape of human destiny. Our shaping tools were words which exposed new facets of understanding about machines, culture, communications, sexuality, life sciences and the like. We see an arrow starting to take form, which suggests a higher meaning to man's existence. And the arrow is pointing to the stars.

The phrases of our common language provide a clue to our destiny. We daily speak of "reaching for a star" when we mean our ultimate goal. Ambition is "reaching for the stars." It is as though we can feel the direction of the arrow pointing, even though we cannot clearly see it.

The pull of destiny's arrow was felt by the whole population of the world when the first human being stepped on the moon. In every corner of the earth, people stopped, interrupted whatever they were doing, to listen or see with hypnotic awe the reality of a human stepping upon the surface of another heavenly body. For the instant that Apollo 11 Commander Neil Armstrong stepped upon the surface of the moon, he lived at the tip of creation's arrow. All mankind felt a part of this adventure, and their hopes and prayers were as one with this lone astronaut. All the earth's leaders and their people became as one in their hope for the success of the mission and for the safety of the astronauts.

The real priorities in the minds of earth's people were shown in the near tragedy of the mission of Apollo 13. All the earth's attention was turned to the struggle for three astronauts to stay alive for their precarious journey back to earth from the moon in their damaged spaceship. During those same moments, hundreds of fine men were dying in the pursuit of their everyday jobs on earth, in automobile, airplane and boating accidents, or as part of their military duty. But the dying of men on earth was amost forgotten, as full attention was turned to the danger facing three men in their tiny, fragile craft.

Our hopes are with those who are truly reaching for the stars. For with them go the hopes of all the human race and the hopes of the life system which created us. It is in the heavens that we look for life's eternity, among other stars and in the dust of stars not born.

AFTERTHOUGHTS

It has been more than two years since the first rough manuscript for this book was completed. In these past two years, there has been time to reflect further on the daily happenings in the world around us. These, then, are thoughts that might occur to anyone who had accepted the thesis of the fourth kingdom two years ago. The fourth kingdom is not a personal thesis. I have simply been another observer of the puzzle, as the pieces were fitted and the picture took form. Having been so close to the subject for seven years, I no longer concern myself with why humanity arose in nature or where we are going. Civilization and the works of society now seem divinely purposeful.

The Manifest Destiny of China

The emergence of China as a potential supernation has been in keeping with the thesis of the fourth kingdom. As the news reporters who accompanied former President Nixon to China in 1973 found out, technological development has top priority in China's affairs. And as many long-term China watchers have observed, the Korean conflicts and the long-term unrest in Indo-China are the results of a policy of China to expand its political and economic influence to those societies which are not already a part of a supernation. China is repeating the manifest destiny saga of the United States and the U.S.S.R. China intends to become an important and powerful custodian of the fourth kingdom.

This all seems appropriate for a people whose ancestors were the inventors and practitioners of the first rockets on earth, the first to invent and employ an astronomical clock, and the leading astronomers of the Middle Ages. Ironically, Chinese astronomers of the eleventh century made the most reliable observations of the spectacular explosive death of a star, a supernova, the remains of which now constitute the Crab nebula.

Whistling in Dark Space

Those of us who were fans of Buck Rogers and Flash Gordon as boys and who watched Star Trek as men have come to feel there is alien life lurking around every constellation in the night sky.

Actually however, we are less certain than ever about the presence of other life in our galaxy. The first astronauts to return from a walk on the moon were placed in quarantine lest they harbor some new germ which could cause a dread disease here on earth. But there was nothing—no sign of life from the moon—not even anything resembling our simplest virus. We were so fearful—or was it hopeful in a negative way?—to assume that the incredible complexity of life could arise in such a hostile environment. We found the moon barren.

We had imagined the planet Venus to be a garden of paradise teeming with tropical life. We envisioned a civilization on Mars that had criss-crossed its planet with huge canals to distribute water from the ice caps. Then our instruments and cameras gave us reality. Venus is a fiery hot planet whose temperature is hundreds of degrees above that of boiling water and whose conditions could not support life. Venus is not a hospitable place where the delicate creation of life could have gone on over these billions of years. Mars appears lifeless; at least no higher forms of life have left significant marks on the landscape. But more than that, Mars contains precious little water and scant amounts of oxygen. It is a dry, dusty planet not conducive to a rich evolution of life. So great doubts about other life in our solar system begin to creep into our reason.

We were certain that other life must be out "there," that we could not be alone in this galaxy of stars. We built an enormous radio telescope that filled a whole valley in the hills near Arecibo, Puerto Rico. We turned it on and listened. Precisely timed bursts of radio waves were picked up in 1967, but these turned out to be natural radiation coming from rotating stars, called pulsars. Years have gone by, and we have heard nothing from other space civilizations. Our galaxy seems hauntingly empty and quiet.

Now, very thoughtfully suppose that earth is the only place where a living cell evolved in all of the Milky Way galaxy where billions of stars shine with the light of life. Let that real possibility slowly register in your living consciousness. Think, then, of the heinous crime of omission, the monumental malfeasance in the office of our very existence, if this society allows the result of billions of years of tedious experimentation to perish.

Let us also suppose that we could some day establish with absolute certainty that another life system existed elsewhere in our galaxy. Would this diminish the urgency of the human mission? A dandelion does not cease seeding the area around it because there is a milkweed in the same meadow. A lion does not cease breeding because there are already leopards or hyenas or zebra on the African continent. The certainty of some alien form of life somewhere has no bearing on the urgent desire of the life system of our origin to find its own eternity among the stars.

Sometime soon, perhaps as you read this book, an unmanned space craft will land on Mars. It will be thoroughly sterilized before leaving, so it will not contaminate Mars with earth's life. The purpose of the landing will be to see whether life has evolved on Mars. This has academic value and is worth one space shot.

But I hope that a little mouse crawls aboard the second lander after it is sterilized and before it is launched. And I hope the mouse had eaten lichen as its last meal and is carrying all sorts of bacteria, spores and seeds in its fur and digestive system.

Then, perhaps a plant kingdom would arise on Mars which would be adapted to the harsh environment. Who knows, some millions of years hence, an animal kingdom will evolve from the plant kingdom, as it did on earth, an animal kingdom which could live comfortably on Mars.

What Happened?

How can society be putting so little effort into the most important mission of human existence? How have we been able to ignore the immense evolutionary pressures to fill lifeless voids in the universe with life? Why haven't we been awakened by the extraordinary devotion and accomplishments by the thousands who worked in our space program?

Why would not all humanity gather around the greatest technological achievement of all mankind, the first visit by earth's life to another planet? Why has this beginning of man's real destiny been allowed to languish because of other "national priorities"?

A part of the problem seems to be that technology alone cannot maintain an orderly civilization. Other skills, "soft sciences," are needed which are critically important in making mankind's path toward higher technology a smoother and a more certain one. A technological society requires the skills of politics, social sciences, law, reporting, finance and many others. Achievements in the "hard sciences" have supported all humanity well and have provided a comfortable life to all members of a technical society. However, men have been enjoying the luxury of technical affluence while forgetting the fact that technology made it all possible.

Thus, there are those in the "soft sciences" who only see civilization's purpose as serving their particular cause.

There are those who visualize the earth as a huge turkey farm dense with people living without worry or care, with their eyes and minds glazed over from the lack of the challenge for which their brains evolved. The welfare budget of the U.S., many multiples of the budget of our space program, is mute testimony to the success of one social cause. Instead of people being caught up in the creative intensity that burned through everyone in the Apollo program, millions are turning to drugs and bizarre behavior to dull their unused minds.

The "hard sciences" also are guilty of diversion of purpose. Many believe all our technology is solely for the welfare and benefit of mankind; that the only value of the space program is the "fallout" which will assist man's life here on earth. They do not see plurality's arrow which points to space. It is beyond some scientific imaginations to think for one millisecond that our huge technology network on earth is nature's way of evolving better space techniques. They cannot see the obvious working of evolution to develop new techniques and new forms by the application of huge numbers of nearly identical trials.

The economist simply does not have a formula that puts a value on technology when used to make possible greater missions in space.

Materialism Without Purpose

Mankind's insatiable appetite for material things stems from an instinctive desire to pursue technology, which in turn drives civilizations to a frenzy of activity. However, without a cause or a purpose, the rush and hurry in uncertain directions to uncertain places creates an excess of technological gimmickry.

Perhaps this continuing quest for more material goods would be less anxious if the cause of this obsession in mankind were universally recognized. If we saw the ultimate use of technology as an extension of nature, with a purpose for the whole life system, perhaps a new life style would evolve. We would see a creative, natural, instinctively satisfying outlet for our energies, and we might all collectively obtain more peace of mind. The waste of technological gimmickry would then disappear.

Hard reality, however, will extinguish our relentless desire for material things, if we do not correct the situation ourselves. We will simply run out of resources and power if our technological explosion continues, as blind as a raging torrent of water flowing in any direction gravity takes it.

American Indian Culture and Values

I visited the Saginaw Chippewa Tribal Community Center recently and saw a four-year old Indian boy playing alone in the day care room. With many toys and activities to choose from, he was studiously putting together a toy bridge and placing toy automobiles on it. He was a symbol of the many Indian people who are willing to meet the world of technology halfway in order to enjoy the comforts and economic benefits of embracing the fourth kingdom.

However, it may be that we children of the Industrial Revolution might want to meet the traditional Indian values halfway in order to improve our outlook and our mental, if not physical, well-being. Over thousands of years, the North American Indians had developed a culture and a system that was in harmony with, yet in awe of, nature. The Indians had a philosophy of life we might look to when we recognize humanity's role of service to nature's purposes, and when the exaltation of mankind that we have been savoring becomes ashes in our mouths.

Traditional Indian culture included well organized and systematic beliefs in the forces of nature and the universe, and an acceptance of unexplainable happenings, which they attributed to the supernatural realm in proper context. They did not separate science from religion. Most importantly, the Indians so well understood the wonders and the power of nature that they did not hold to any delusion that man was exalted.

It may be time to find and resurrect more of the Indian's basic philosophy. It will help us all to place technology, materialism and nature in the proper perspective.

A New Book

I wish the book, *The Lives of a Cell,* by Lewis Thomas, had come out years ago. It would have given me determination to publish this book earlier. The oneness and interdependence of all earth's life system are a required premise in order that the thesis of the fourth kingdom stand up to its critics. While there has been a smattering of evidence to that effect, it has never been so succinctly said or so clearly articulated as in *The Lives of a Cell.*

Golden Vision

My son left one of his tapes in our stereo player one day, and I was inadvertently exposed to Neil Young's "After the Gold Rush." The words and thoughts in the song were imprecise in the manner of an oracle speaking. But the song told of seeing, "in a dream," the loading of a space ship. It spoke of nature's seed which would carry life to a new home, perhaps the sun.

AFTERTHOUGHTS

I had concluded long ago that humanity's passion for gold, the gold fever, the gold rushes and the hoarding of gold were related to the future needs for great quantities of this metal on a space vehicle. Our Apollo program showed how gold was a critical requirement for a space ship.

Apollo-Soyouz

Through this joint space mission, involving Apollo of the United States and Soyouz of the U.S.S.R., mankind took another step toward fulfilling humanity's biological reason for existence. This momentous event symbolized what we must all do together if the living earth is to survive.

Two of the supernations of earth joined together to demonstrate the highest possible level of achievement by mankind and his creations of the fourth kingdom.

Hopefully, the new era is beginning now, where peoples of the earth, regardless of their race, creed, or political ideologies, will contribute their share of creativity to humanity's mission to the stars and, in return, share in the glory of the realizations and the rewards.

BIBLIOGRAPHY

A thesis which derives its origins from all history and technology must be anchored to the most recently developed findings of science and an accurate view of history. I am indebted therefore to the contributors, authors, editors, and publishers of books that accurately summarize great quantities of information, especially to those who made possible the *Encyclopaedia Britannica* and the Time-Life Books' *Nature Library* and *Science Library*. These books are based on countless references. Without benefit of their clear, concise analyses, research to relate all facets of the fourth kingdom to the knowledge of the world would have taken more than one lifetime.

Alexander, Tom. "The Costly Hunt for the Heart of Matter." *Fortune,* (March, 1968).

Alexander, Tom. "The Hot New Promise of Thermonuclear Power." *Fortune,* (June, 1970).

Alexander, Tom. "The Shimmery New Image of Matter." *Fortune,* (June, 1968).

Alexander, Tom. "The Unexpected Payoff of Project Apollo." *Fortune,* (July, 1969).

Allen, D. L. and Mech, L. D. "Wolves versus Moose on Isle Royale." *National Geographic,* (February, 1963).

Ardrey, Robert. *Territorial Imperative.* New York: Atheneum, 1966.

Beiser, Arthur and the Editors of Time-Life Books. *The Earth.* New York: Time-Life Books, 1971.

Bergamini, David and the Editors of Time-Life Books. *The Universe.* New York: Time-Life Books, 1970.

Brown, Dee. *Bury My Heart at Wounded Knee.* New York: Holt, Rinehart and Winston, 1971.

Calvin, Melvin. *Chemical Evolution.* New York and Oxford: Oxford University Press, 1969.

Confraternity of Christian Doctrine. *The New American Bible.* Saint Joseph Edition. New York: Catholic Book Publishing Company, 1970.

Darwin, Charles. *The Origin of Species.* (1859). London: Everyman's Library, Dent & Sons, 1956.

Encyclopaedia Britannica. Chicago: William Bentor, 1963.

Engel, Leonard and the Editors of Time-Life's Books. *The Sea.* New York: Time-Life Books, 1971.

Gibran, Kahlil. *The Prophet.* New York: Alfred A. Knopf, 1968.

Goddard, Robert H. "The Ultimate Migration." Manuscript dated January 14, 1918. The Goddard Biblio Log, Friends of the Goddard Library, November, 1972.

Goudsmit, Samuel A., Claiborne, Robert and the Editors of Time-Life Books. *Time.* New York: Time-Life Books, 1971.

Howell, F. Clark and the Editors of Time-Life Books. *Early Man.* New York: Time-Life Books, 1971.

Hassrick, Royal B. *The Sioux, Life and Customs of a Warrier Society.* Norman: University of Oklahoma Press, 1964.

Hefley, James C. "In the Beginning, God . . ." *Grit,* (December 27, 1970).

Holy Bible. King James Version. New York and London: Oxford University Press.

McKenna, Richard. *The Sand Pebbles.* New York: Harper & Row, 1962.

Michener, James A. *Hawaii.* New York: Random House, 1959.

Momaday, N. Scott. "A Vision Beyond Time and Place." *Life,* (July 2, 1971).

Morris, Desmond, *The Naked Ape.* New York: McGraw Hill, 1967.

Pfeiffer, John and the Editors of Time-Life Books. *The Cell.* New York: Time Life Books, 1970.

Pyle, Ernest T. *Here is Your War.* New York: Holt, 1943.

Salkeld, Robert. "Space Colonization Now?" *Astronautics and Aeronautics,* (September, 1975).

Seaborg, Glenn T. et al. "Research—the Painful Struggle for Relevancy." *Business Week,* (January 2, 1971).

Storer, Tracy Irwin and Usinger, Robert L. *General Zoology.* New York: McGraw Hill, 1965.

Teilhard de Chardin, Pierre. *The Phenomenon of Man.* New York: Harper and Row, 1965.

Thomas, Lewis. *The Lives of a Cell.* New York: Viking, 1974.

Watson, Lyall. *Supernature.* New York, Doubleday, 1973.

Woodwell, George M. "The Energy Cycle of the Biosphere." *Scientific American,* (September, 1970).

Young, Richard S., Klein, Harold P. et al. "NASA Studies Planetary Habitation Methods." *Aviation Week and Space Technology,* (November 30, 1970).